Single Sourcing

Building Modular Documentation

Single Sourcing

Building Modular Documentation

Kurt Ament

William Andrew Publishing
Norwich, NewYork, U.S.A.

Library or Congress Catalog Card Number: 2002026287

ISBN: 0-8155-1491-3
Printed in the United States of America

Published in the United States of America by
Noyes Publications / William Andrew Publishing
13 Eaton Avenue
Norwich, NY 13815
1-800-932-7045
www.williamandrew.com
www.knovel.com

10 9 8 7 6 5 4

This book may be purchased in quantity discounts for education, business, or sales promotional use by contacting the Publisher.

Library of Congress Cataloging-in-Publication Data
Ament, Kurt, 1960-
Single sourcing: building modular documentation / Kurt Ament.
p. cm.
Includes index.
ISBN 0-8155-1491-3 (alk. paper)
1. Technical writing. I. Title.
T11 .A39 2002
808'.0666--dc21 2002026287

CONTENTS

FOREWORD

For over a decade, I have been doing what is now called single sourcing. Three experiences, in particular, have shaped the way I have presented single sourcing in this guide.

In the early 1990s, I worked as a senior technical writer for a large defense company in Southern California. There my publications group developed internal documentation with a mainframe-based publishing system written in the Standard Generalized Markup Language (SGML). One day, our division announced it was "going paperless" within one year. To meet this hard deadline, we leveraged the power of SGML to convert all print documentation to online documentation. To enable users to access large quantities of information online, we developed strict syntax guidelines for headings and index entries. To index a large volume of documents with very limited resources, we developed a sophisticated script that generated indexes automatically, based on our syntax guidelines. The new navigation system worked remarkably well. Our users told us they were able to find the information they needed without the benefit of printed pages.

In the mid-1990s, I worked as a senior technical editor for a multinational software company, also in Southern California. One day, my publications group was told that our flagship product was going to be completely redesigned for a new operating system. No one had yet seen the new operating system. Nevertheless, our product and the new operating system were scheduled for release on the same day. Naturally, the documentation had to be ready as well. To meet this seemingly impossible deadline, we threw the entire weight of our team behind the online help version of our documentation, rather than splitting the team into print and online subteams, as had previously been the case. We then hammered out a document structure, based on the limitations of our online help format. Once we had completely restructured and rewritten the documentation, we converted our modules into a standard template for our printed user guide. The guide that resulted was more usable than its linear predecessors.

The Society for Technical Communication gave our guide the highest award in its category, calling it: "well-written and professional, clearly an excellent example of user-friendly technical communication that represents the state-of-the-art and the best in our profession."

In the late 1990s, I worked as a user interface designer for one of the largest multimedia companies in Germany. There I was part of a small team that built what was originally intended to be a companion website for the first nationwide financial newspaper to be founded in half a century. Our client, the founder of the newspaper, selected a high-end content management system that leveraged the Extensible Markup Language (XML). In the space of a few months, we built modular online templates that enabled editors to publish their print articles online in real time. Users could view the same articles in print or on the Internet, or download up-to-the-minute summaries to handheld devices. Shortly after the launch of the newspaper, I presented a morning edition, along with its Internet and handheld versions, to colleagues at a usability seminar in the Netherlands cosponsored by the Nielsen Norman Group. What seemed to impress my colleagues most was not the high-tech bells and whistles but the fact that the same stories appeared in all three media, and that all three "felt" like a newspaper. My colleagues seemed especially delighted that the structure of the online versions mirrored the major sections of the newspaper itself.

My experiences in the defense, software, and multimedia industries have taught me three basic lessons about single sourcing. First, single sourcing is a methodology, not a technology. Although single sourcing can involve very complex technologies, modular writing is what drives the process. If you do not develop modular content, no tool or technology can transform your content into usable documentation. But if you do develop modular content, document assembly is relatively easy. And usable and re-usable documentation is almost a foregone conclusion.

Second, modular documentation is not possible without standardization. Writing standards that are "nice to have" in linear documentation become "must haves" in single sourcing. To develop modular content that works "anywhere, anytime, anyhow," you need writing guidelines that answer usability and re-usability questions directly. Ultimately, the only way to answer usability and re-usability questions is by trial and error. Only experience tells you what works, what does not work, and why. If you base your writing guidelines on live projects with real deadlines, you can re-use your experiences in future projects.

Finally, modular writing guidelines developed for single sourcing projects really do improve the quality of your documentation. These guidelines make usability *the* central issue. Even if you are not currently single sourcing, you can leverage guidelines developed for single sourcing projects to improve the usability of your traditional projects.

You will notice that this guide makes repeated references to Adobe FrameMaker and Quadralay WebWorks Publisher. These references are not accidental. In the past year, I have extensively customized WebWorks Publisher templates for a large multinational corporation, which now uses them to convert FrameMaker+SGML documents to online help. Instead of describing single sourcing abstractly, I decided to reference these two tools, which are so widely used by technical writers, to help draw a realistic picture of the single sourcing process. Your own experience will vary.

Kurt Ament

June 2002

About this guide

Single sourcing is a method for developing re-usable information. Although project planning and document conversion are essential to any single sourcing project, it is modular writing that ultimately determines success. This guide explains in plain language and by example how to build modular documentation you can re-use in different formats for different audiences and purposes.

This guide begins with a conceptual overview of the single sourcing method. It then walks you through a 10-step process for converting linear documentation to modular documentation. Each step of the process is cross-referenced to detailed guidelines. These guidelines show you by example how to structure content, configure language, and leverage technology to maximize usability and re-usability.

To provide you with a realistic picture of the single sourcing process, the steps, guidelines, and examples in this guide are very specific. Each is based on successful single sourcing projects in the real world. To provide you with a flexible framework you can re-use for your own purposes, the guidelines are modular. Pick the guidelines you need. Then customize them to meet your specific needs.

Audience

This guide is designed for anyone interested in single sourcing:

- **Technical writers and editors**

  The guide benefits technical writers and editors who want a flexible framework for developing successful single-source documentation.

- **Technical publications managers**

  The guide benefits technical publications managers who want to standardize corporate publications in a way that improves the quality of print and online documentation while saving time and money.

- **Teachers and students of technical communication**

  The book benefits teachers and students of technical communication who want to learn more about single sourcing as it is actually practiced in corporate publishing environments.

Even if you are not involved in single sourcing, this guide provides you with guidelines to improve the usability of your documentation.

Purpose

This guide has two integrated goals:

- **Immediate success**

  The primary goal of the guide is to show technical writers how to develop successful single-source documentation. The guide presents modular writing techniques that have been proven in real-world publishing environments. Rather than explaining what should work, the guide explains what does work, what does not work, and why.

- **Continuing success**

  The secondary goal of the guide is to provide technical publishing groups with flexible single sourcing guidelines. This framework can be customized by these groups to fit the specific requirements of their companies, products, and customers.

This guide provides usability and re-usability guidelines you can integrate into your in-house documentation standards.

Organization

This guide is organized into the following chapters:

- **Chapter 1. "About single sourcing"**

  Conceptual overview explains what single sourcing is, reasons for single sourcing, how single sourcing works, types of single sourcing, and how to implement single sourcing successfully.

- **Chapter 2. "Building documents"**

  Process explains how to transform linear documentation into modular documentation. Each of the 10 steps in the process is cross-referenced to detailed writing guidelines. The process also explains how to re-use your experiences to build project style guides.

- **Chapter 3. "Structuring content"**

  Guidelines explain how to build, assemble, and link modular content. The guidelines explain how to write content modules, how to organize modules into documents, and how to cross-reference the documents.

- **Chapter 4. "Configuring language"**

  Guidelines explain how to optimize language for online documents. The guidelines also explain how to maintain a single, user-centered writing style while developing documents with multiple authors.

- **Chapter 5. "Leveraging technology"**

  Guidelines explain how to leverage industry-standard tools and technologies to save time and money. The guidelines also explain how you can use these tools to automate your documentation standards.

Each group of guidelines is organized alphabetically, for easy reference. The guidelines include negative and positive examples, as well as cross-references to related guidelines.

This guide includes a glossary and an index.

CHAPTER 1

About single sourcing

Single sourcing is a documentation method that enables you to re-use the information you develop. You develop modular content, then assemble that content into different formats, such as printed manuals, online help systems, even websites. Re-using information saves you time and money because it eliminates duplicate work.

Single sourcing also increases the usability of your documentation. By developing modular information that is usable in any format, you raise your documentation standards. Usability standards that are "nice to have" in traditional documents become "must haves" in modular documents. In effect, single sourcing forces you to do the right thing for your users.

Single sourcing is a methodology, not a technology. Although the software tools associated with single sourcing are complex, it is modular writing, not technology, that ultimately determines the success of your single sourcing projects. To ensure success, develop local, project-based standards for modular writing. Base your standards on what actually works in your own projects.

In this chapter

This chapter contains the following sections:

What is single sourcing? 3

Single sourcing is a method for re-using information. To re-use information, single sourcing separates input from output in the documentation process.

Reasons for single sourcing 8

Single sourcing saves time and money, improves the usability of your documentation, and increases team synergy.

How single sourcing works 11

In single sourcing, you build modular content, assemble the modules into documents, and link the modules together.

Types of single sourcing 15

You can use single sourcing to deliver the same modular content in different document formats. Or you can use single sourcing to re-organize modular content for different document types.

Successful single sourcing 19

To succeed at single sourcing, start with small but live test projects. Separate information architecture and information development. And establish success-based guidelines for modular writing.

TIP For information about single sourcing development tools and technologies, see Chapter 5, "Leveraging technology."

What is single sourcing?

Single sourcing is a method for systematically re-using information. With this method, you develop modular content in one source document or database, then assemble the content into different document formats for different audiences and purposes.

Single sourcing changes the way you develop information:

- **Re-usable content**
  Unlike format-based documentation, content-based documentation can be re-used in different formats. Single sourcing separates content (input) from format (output) in the information development process. You concentrate on developing content, not format.
- **Modular writing**
  Unlike linear (document-based) writing, modular (element-based) writing can be re-used in different document formats. You write stand-alone modules, not documents tied to a particular format.
- **Assembled documents**
  Unlike traditional desktop publishing, single sourcing assembles different documents from the same content. You use conversion tools to convert modular content into different document formats.

Although its primary goal is to save time and money, single sourcing improves the quality of your documentation. By requiring you to develop modular information that is usable in any format, single sourcing makes document usability an all-or-nothing issue. If your content is modular, your single sourcing project succeeds. If not, it fails. To succeed, you have no choice but to incorporate usability guidelines into your documentation standards. Although usability is a goal for most technical writers, single sourcing makes it the top priority. In effect, single sourcing forces you to implement the usability standards you have always talked about.

TIP For more about the relationship between single sourcing and document usability, see "Improving document usability" on page 9.

About re-usable content

With the single sourcing method, you concentrate on content, not format. Content-based documentation is very different from traditional, format-based documentation, which is written for a particular document format, and is usually designed to be read in a given sequence.

Format-based content is not re-usable

In the traditional documentation process, you hand-craft individual documents for a particular output format, such as the following:

- Online help
- Printed manual
- Training manual
- Website

Because content is tied to a given format, it cannot be easily re-used in a different format. At best, re-use involves extensive rewriting.

Content-based content is re-usable

In the single sourcing process, you separate content (input) from format (output) when developing information. Typically, you use content-based information development tools that leverage the Standard Generalized Markup Language (SGML) or the Extended Markup Language (XML).

Content-based development tools enable you to develop information at the element level rather than the document level:

- **Document level (linear)**
  In traditional publishing, you develop linear content (for example, a printed manual designed to be read cover to cover).
- **Element level (modular)**
  In single sourcing, you develop modular content (for example, sections, paragraphs, and lists that make sense in any context).

If you develop modular content that is not tied to any given format, you can re-use the content in many different formats.

TIP For related guidelines, see "Development tools" on page 188.

About modular writing

Modular writing is the opposite of linear writing. Linear writing assumes a particular document format, and a particular reading sequence. In stark contrast, modular writing assumes nothing. You build stand-alone content modules that make sense in any document format or reading sequence.

Linear writing is not re-usable

Linear writing is hierarchical and sequential:

- **Hierarchical**

  Linear writing organizes information into hierarchies, based on the structure of the document itself. For example, a printed manual might cluster all conceptual, procedural, and reference information related to the installation of a given product in a chapter called "Installation."

  Such a thematic hierarchy mixes three different types of information:

  - Topics
  - Procedures
  - Definitions

  Before you could re-use the information in another document, you would have to separate the different types of information. To do so, you would have to restructure the information.

- **Sequential**

  Linear writing assumes a given reading sequence. It is intended to be read from beginning to end. For example, a chapter in a printed manual might include installation instructions before configuration instructions. The installation section of the chapter might include a note pointing to configuration instructions "later in this chapter." Likewise, the configuration section might include a note pointing to installation instructions "earlier in this chapter." Obviously, these sequential references would not make sense online. Before you could re-use the information in an online help system, you would have to remove all references to sequence. To do so, you would have to rewrite the information.

Linear documentation is almost impossible to re-use.

Modular writing is re-usable

In contrast to linear writing, modular writing is non-hierarchical and non-sequential. You build stand-alone content modules that make sense in any context. The structure and sequence of modules is determined by different documents, which you configure and assemble separately.

Modular writing involves chunking, labeling, and linking:

- **Chunking (content)**
  In modular writing, you "chunk" information into stand-alone modules, based on the type of information being presented. For example, you might use descriptive text to explain what a product is and how it works. You might use procedures to explain how to operate the product. You might use definition lists to describe product menus and options. You might use flowcharts to illustrate product processes. And you might use tables to list complex installation requirements.
- **Labeling (headings)**
  In modular writing, you use standardized, context-independent syntax to label information chunks so they make sense no matter where they appear. For example, you might label superprocedures with gerunds (for example, "Printing") and subprocedures with infinitives (for example, "To print a document to a file").
- **Linking (cross-references)**
  In modular writing, you link labeled chunks with cross-references. For example, you might link each step in a superprocedure to its related subprocedure. You might link each subprocedure to related topics, procedures, and definition lists. And you might generate a unique table of contents and index for each assembled document.

By chunking, labeling, and linking information in this way, you develop content that is relatively easy to assemble into different documents.

TIP For related topics, see the following sections:

- "Building modular content" on page 12
- "Linking modular content" on page 14

About document assembly

Once you develop stand-alone content modules, you assemble them into different document formats for different audiences and purposes:

- **Audiences**

  You can use conditional text to include or exclude sections for novice or expert users, system administrators or operators, and so on.

- **Purposes**

  You can configure the hierarchy and sequence of modules for concept-driven user guides, task-driven online help, and so on.

- **Formats**

  You can automatically generate printed manuals, online help systems, training materials, websites, and so on.

Document assembly does not change the content modules. It simply determines the order and format of the content modules.

TIP For some writers, "assembly" has negative connotations. The word conjures images of assembly line workers doing drone-like work. The fear is that document assembly removes the last vestige of creativity from technical writing. Nothing could be further from the truth. By eliminating wasteful and meaningless duplication of effort, document assembly removes a tremendous amount of drudgery from technical writing.

Document assembly enables writers to do what they do best. Writers who care most about information development can focus on document content without wrestling with publishing technologies. And writers who care most about information architecture can configure document development tools to generate new documents in different formats quickly and easily.

For related topics, see the following sections:

- "Assembling documents" on page 13
- "Organizing smart teams" on page 21

Reasons for single sourcing

There are three basic reasons for single sourcing:

- **Saving time and money**

  By re-using the same content in different document formats, you reduce duplicate work and increase flexibility.

- **Improving document usability**

  By developing modular content that is usable in any format, you improve the usability of your documents in all formats.

- **Increasing team synergy**

  By developing shared modules, you ensure wise consensual decisions about everything from style guidelines to document templates.

Single sourcing changes the way you do business, for the better.

Saving time and money

When properly implemented, single sourcing saves you time and money. Single sourcing enables you to develop content once, then re-use that content in many different ways. By re-using the same content in different formats for different audiences and purposes, you reduce duplicate work. For example, if you develop a printed manual and an online help system for one product, it is very likely that the two documents contain very similar information, albeit in different formats. If you develop the two documents separately, you do the same work twice. But if you develop re-usable content modules from a single source, you do the same work only once. In other words, you cut your workload in half.

Single sourcing does more than reduce your workload. It increases your flexibility. It is not uncommon for document requirements to change after a project has begun. For example, at the beginning of a project, you might be told to develop an online help system for a given product. Then, when you are halfway through the project, you might be told the product team wants some way to print out the entire online help system in a format that is easy to review, page by page. To meet this new requirement, you would most likely have to produce a print-ready electronic book. If you are already single sourcing, converting your content to this second format is relatively easy. Single sourcing prepares you for the unexpected.

TIP To find out how to cut localization costs, see "Localization" on page 195.

Improving document usability

At its best, single sourcing forces you to develop information that is usable in any situation. In other words, *usability is a prerequisite for single sourcing*. This all-or-nothing prerequisite raises usability standards exponentially. The same content has to be usable when accessed from an index in a printed manual, from a link in an online help system, or from a search engine on a website. By forcing you to develop content that is usable in any situation, *single sourcing forces you to develop re-usable documentation*.

To develop information that is usable in any format, you plan for the worst-case scenario:

- **Print (best case)**

  Because they have had centuries to adapt to the printed page, people have a high tolerance for poorly constructed print documents. For example, if users cannot find what they are looking for in the table of contents or index of a printed manual, they often flip through the manual, quickly scanning the text for information that "pops out" at them. That is, users abandon the navigation scheme of the manual and follow their own "inner radar."

- **Online (worst case)**

  Because they have had only a few decades to adapt to online media, people have a low tolerance for poorly constructed online documents. If users cannot find what they are looking for online, they simply get lost. If you lose users, you lose customers.

By developing information for online access rather than print access, you not only ensure the information is usable online, you increase the usability of the information in any format.

TIP Anyone who has ever converted online documents to a print format has had an "aha" experience. What worked well online worked even better on paper.

Even if you are not ready for single sourcing, you can improve your print documentation by thinking "online help" when developing information. Does the information on your computer screen make sense when taken completely out of context? If not, rewrite the information.

Increasing team synergy

Documentation is a collaborative effort. Although individual documents are theoretically "owned" by individual writers, these documents are actually the result of a team effort between product developers, template developers, and other information developers, such as writers, editors, and indexers. Responsibility is collective.

By separating content from format in the documentation process, single sourcing makes it clear to individual team members that they are developing parts of the whole, not the whole itself. The more sophisticated the single sourcing process, the more team members are allowed to concentrate on what they do best (for example, building, assembly, or linking). In the process, team members become more interdependent.

By making team members more interdependent, single sourcing acts as a catalyst for team synergy. For example, single sourcing is a pragmatic basis for consensual group decisions about writing guidelines and document templates. What are often subjective, theoretical, and optional guidelines for writing style and document format become objective, practical, and necessary. For single sourcing to work at all, the team needs to make hard decisions about many issues, big and small. Once everyone understands that they sink or swim together, they rise to the occasion, making smart decisions that ensure success.

NOTE Good publishing teams sometimes make bad decisions. But publishing teams whose members share responsibility for the success or failure of each other tend to correct their mistakes quickly and permanently.

Single sourcing raises the stakes for team decisions. Bad decisions hurt the team. Good decisions help the team. By holding each team member accountable for the success of the team, single sourcing provides teams with a premise on which all team members can agree: survival. Because team failure threatens individual survival, success becomes the only option. In its effort to survive, the team *chooses* success.

How single sourcing works

With single sourcing, you develop re-usable information in one source document or database. You build modular content in this single source, then assemble the content into document formats that target specific audiences and purposes.

In the most general terms, single sourcing is a three-step process:

1. **Building**
   You build modular content (for example, topics, procedures, definitions, figures, tables, and so on) that answer specific questions. These modules make sense even when taken out of context.
2. **Assembling**
   You organize content modules into different types of documents (for example, printed manuals, online help systems, training manuals, websites, and so on). Each assembled document has a distinct audience, purpose, and format.
3. **Linking**
   You build cross-references (for example, tables of contents, section contents, inline cross-references, indexes) that link modules logically for a given document. These cognitive bridges connect stand-alone modules, transforming them into coherent documents.

The single sourcing process is often cyclical. For example, you might decide to re-organize a document after seeing what it looks like in a given output format (for example, online help). Once you re-organize the document, you can then fine-tune the links. And so on.

TIP Although cyclical single sourcing sounds inefficient, it improves the quality of documents. By progressively synchronizing your input (content) and output (format), you perform real-time quality assurance.

Building modular content

To develop re-usable information, you build small, focused, stand-alone content modules that make sense even when taken out of context.

Types of questions

Content modules answer basic questions:

- Who?
- What?
- When?
- Where?
- Why?
- How?

As a rule, each content module answers one question only.

Types of answers

To answer specific questions, you develop specific types of content modules, such as the following:

- **Topics**

  Descriptive texts that explain who, what, when, where, and why. Such texts are especially useful when discussing concepts.
- **Procedures**

  Step-by-step instructions that explain how to do something. These instructions can be superprocedures, subprocedures, and so on.
- **Definitions**

  Simple or complex lists that define terms. These lists can explain menu items, commands, technical terms, and so on.
- **Figures**

  Images that depict physical objects, conceptual processes, and so on. Common figures are screenshots and flowcharts.
- **Tables**

  Complex, two-dimensional lists that provide an overview of topics that are difficult to explain with text alone.

Assembling documents

Once you have developed stand-alone content modules, you assemble the content modules into documents.

Modular content can be assembled into a wide variety of document types, such as the following:

- **Printed manuals**
  Documents designed to be printed (for example, reference manuals, user guides, and so on). These documents can be bound books or print-ready electronic books.
- **Online help**
  Documents designed in any number of different online help formats (for example, Microsoft HTML Help, Microsoft WinHelp, Oracle Help for Java, Sun JavaHelp, and so on). These documents can be context-sensitive help or online documentation.
- **Training materials**
  Documents designed for training sessions (for example, slide shows, workbooks, and so on). These documents can appear in a variety of print and online formats.
- **Websites**
  HTML-based documents displayed as webpages on the Internet. These documents should not be confused with HTML-based online documentation, which consists of webpages viewed locally by users.

These document types do not change the content modules. They simply determine in which order and in what format the modules are presented.

TIP How you assemble single-source content modules into documents is dependent on the publishing software you use. For example, you may decide to assemble a document after linking its modules.

For more about publishing software, see "Development tools" on page 188.

Linking modular content

As a rule, users scan documents for specific information, rather than reading them from cover to cover (print), or from start to finish (online). When linking modular content, your goal is to get users the information they need as quickly as possible. To help users navigate modular content, you link stand-alone content modules with cognitive bridges, or cross-references. Without extensive cross-referencing, assembled documents are not as user-friendly as custom-built documents.

After assembling documents, you build distinct types of cross-references, such as the following:

- **Tables of contents**

  All assembled documents, be they print or online, require some sort of table of contents. Although a table of contents can take many forms, it always shows the structure of the assembled document.

- **Section contents**

  In major sections of print and online documents, it is very helpful to provide users with "advance organizers," annotated lists that describe the subsections contained in the sections.

- **Inline cross-references**

  When separating different levels of information into distinct content modules, single sourcing physically separates modules that are thematically related. To make the relationships between the related modules explicit, you add inline cross-references that link the modules (for example, from a procedure to a definition list, and vice versa).

- **Indexes**

  To help users locate information quickly and easily, you build document indexes. These alphabetized cross-reference lists show relationships between topics, procedures, and so on.

To test your cross-references, it is a good idea to assemble documents early in the document development cycle.

TIP With a sophisticated authoring tool, you can manually add cross-reference markers to content modules. You can then automatically assemble the cross-references along with their parent documents. For more about authoring tools, see "About authoring tools" on page 189.

Types of single sourcing

There are two types of single sourcing, each with somewhat different goals:

- **Repurposing**

  Delivering the same content in different output formats. Typically, content is developed as a complete document (for example, a printed manual), then converted to another format (for example, online help). You develop content that is embedded, or pre-assembled, in a given document. You then convert the document into different formats. For repurposing, writers often use document-driven publishing tools.

- **Re-assembly**

  Re-organizing modules for different audiences and purposes. Before you convert content modules to another format, you reconfigure them for the new format. Typically, content is *not* developed as part of a complete document. Instead, it is developed at the module level. For re-assembly, writers often use database-driven publishing tools.

Although these two approaches have different strategic goals, the difference is one of degree, not of kind. In practice, all single sourcing projects involve some degree of repurposing *and* re-assembly.

NOTE The distinction between repurposing and re-assembly is a useful fiction that helps you clarify project goals. Think of repurposing as "re-assembly lite." Think of re-assembly as "extreme repurposing."

For information about document- and database-driven tools used in repurposing and re-assembly, see "Development tools" on page 188.

About repurposing

Repurposing is the process of delivering the same content in different formats. As in traditional documentation, you develop content as a complete document (for example, a printed manual). You then convert the complete document to another format (for example, online help).

Repurposing is more than document conversion:

- **Mechanical conversion (format)**

  Document conversion is the process of transforming information from one document format (for example, SGML) to another format (for example, HTML). The goal of document conversion is to ensure *consistency* in different document formats. Because it is a mechanical process, document conversion is best performed by machines.

- **Cognitive repurposing (content)**

  Repurposing is the process of modifying information developed in one document format (for example, a printed manual) so that it makes sense when converted to another format (for example, online help). The goal of repurposing is to ensure the *usability* of identical content in different document formats. Because it is a cognitive process, repurposing is best performed by humans.

Repurposing forces you to develop information that works well in more than one document output format. For example, if you know that the printed manual you are writing will eventually be converted to an online help system, you develop information modules in such a way that they are equally usable in print documents and online documents. By forcing you to develop information that is usable in any format, repurposing forces you to develop re-usable information.

TIP Developing printed manuals for re-use as online help actually improves the quality of your printed manuals. For a discussion of this phenomenon, see "Improving document usability" on page 9.

About re-assembly

The concept of assembly is central to single sourcing. In repurposing, you develop information as content modules embedded, or pre-assembled, in a given document. You then convert these content modules into a different document format. In effect, you assemble the modules in a new format. Re-assembly takes the single sourcing process one step further. Not only do you convert modules to another format, you also re-organize the modules for new audiences and purposes.

Sample re-assembly problem

In a re-assembly project, you might decide to develop information for a printed manual, an online help system, and a training manual from the same source document. Using the same source document for the printed manual and the online help system would likely involve repurposing, but not re-assembly, because both formats share the same audience and purpose. However, if you decide to use the single-source document for a training workbook, you introduce a new audience and purpose.

To accommodate user training in your single-source document, you need to control which modules appear in your printed manual, which appear in your online help system, and which appear in your training workbook. To do so, you need to go beyond repurposing. You need to re-assemble.

You might decide to re-assemble your content modules as three distinct documents:

- **Concept-oriented user guide**

  You might want your printed manual to introduce conceptual information before introducing sequential procedures.

- **Task-oriented online help**

  You might want your online help system to begin with procedures before discussing conceptual information. Also, you might want to list procedures alphabetically for easy reference.

- **Classroom-oriented training workbook**

  You might want your training workbook to introduce users to basic concepts first, then provide them with hands-on exercises.

Sample re-assembly solution

To re-assemble your content as a printed manual, an online help system, and a training workbook, you could use the following process:

1. **Mark source documents.**

   In your single-source document, you can indicate which modules are shared by all three documents, which are for the printed manual only, which are for the online help system only, and which are for the training workbook only.

2. **Customize conversion templates.**

   In your conversion templates, you can indicate which modules should appear in which output documents, and in which order they should appear. That is, you can create three document structures, each with a distinct table of contents and index.

3. **Re-assemble documents.**

   Once your conversion templates are in place, you can re-assemble your three output documents. By re-assembling the three documents early in the development process, you can synchronize input (content modules) and output (formatted documents), thereby performing real-time quality assurance.

A single-source document can be a formatted document, such as a book. For example, your source files for a printed manual might also contain modules for an online help system and a training workbook.

You can electronically mark (tag) sections of the single-source document for the printed manual, online help system, or training workbook. Then you can easily control your output by switching sections on or off. When printing the manual, you can switch off the online help and training sections. When generating the online help, you can switch off the printed manual and training sections. And so on. Of course, there is no need to mark sections shared by all three documents.

TIP High-end authoring and conversion tools enable you to mark document elements with conditional text tags for specific output formats (for example, printed manuals, online help systems, training materials, and so on).

For related guidelines, see the following sections:

- "Conditional text" on page 183
- "Development tools" on page 188

Successful single sourcing

Successful single sourcing is not an accident. It is the result of careful planning, and even more careful implementation. When implementing single sourcing, remember that your basic goal is to save time and money. Like your goal, your implementation should be extremely pragmatic.

Successful single sourcing is local, not global. Although the single sourcing method is universal, your implementation is unique. When implementing single sourcing, do not do what should work in theory. Do what does work for you in practice. What works for you is the only measure of success.

In single sourcing, success depends on three basic factors:

- **Realistic goals**
  Do not try to change the world overnight. Set modest goals for small but real projects with real deliverables and real deadlines.
- **Smart teams**
  Let people do what they do best. Centralize information architecture, decentralize information development, and overcommunicate.
- **Writing guidelines**
  Establish guidelines for modular writing. Base your guidelines on successful online documents rather than print documents.

TIP For guidelines you can re-use in your own single sourcing projects, see Chapter 3, "Structuring content," and Chapter 4, "Configuring language." The writing guidelines in these chapters are based on successful projects. Customize them to meet the realities on the ground where you work.

Setting realistic goals

When you first implement single sourcing, set modest goals for small but live projects with real deliverables and real deadlines. Small projects keep you safe, small steps keep you sane, and live projects keep you honest.

Small projects keep you safe

Small, focused projects and teams provide you with maximum flexibility. When faced with inevitable forks in the road, small teams can reach agreements quickly, and act immediately. Small projects provide you with a psychological edge. If you are working with relatively small deliverables, you are more likely to take intelligent risks. If your experiments go wrong, you can always backtrack and recover. Up-front planning is essential but provisional. You learn your most important lessons through trial and error. If you can experiment freely, you can improve your processes in very tangible ways. You can then re-use these process improvements as best practices for future projects.

Small steps keep you sane

Your first duty in any project is survival. To avoid overextending yourself, take small steps. Set modest goals that can be achieved easily. Once you achieve these initial goals, you can set new, more ambitious goals. By ratcheting up your goals gradually, you meet your deliverables without losing your mind. You can then adjust your goals to the realities on the ground in real time.

Live projects keep you honest

You cannot discover real solutions without encountering real problems. The best place to encounter real problems is in live projects. When you test your single sourcing process on a live project with real deliverables and real deadlines, you are highly motivated to succeed. Failure is not an option. By overcoming problems you encounter along the way, you develop solutions you can then re-use for other projects.

Organizing smart teams

When setting up single sourcing teams, let team members do what they do best. Centralize information architecture, decentralize information development, and encourage overcommunication.

Centralize information architecture

Establish information architects to select, customize, and maintain single sourcing tools that meet the real requirements of real projects in real time. These tools should be stored in a central location under central control. Do not allow information architects to work in an ivory tower. Instead, force them to troubleshoot all technical problems encountered in the field by information developers. Set up a process that enables your architects to incorporate proven solutions to local problems into global templates.

Decentralize information development

Do not force information developers to get under the hood of complex single sourcing applications. Instead, provide them with the tools they need to meet their immediate deliverables. Only by concentrating on modular writing can they develop usable and re-usable documentation. Empower information developers to set up modular writing guidelines. These consensual guidelines should consist of best practices that have proven themselves in real projects. They are your collective memory.

Encourage overcommunication

Information architects think globally. Information developers think locally. For these two groups to understand each other, they have to do more than meet each other halfway. They have to overcommunicate. Actively encourage overcommunication between information architects and information developers. If they work together, you can leverage their very different types of expertise to build airtight publishing solutions.

TIP The best information architects are engineers with the souls of writers. Oftentimes, they are information developers with a strong interest in publishing technologies. You know who they are. Put them to work.

Developing writing guidelines

Modular writing, not programming, ultimately determines the success or failure of single sourcing projects. Before starting a single sourcing project, establish consensual guidelines for modular writing. Base your guidelines on past success. Update the guidelines with new success, as you achieve it.

Establish consensual guidelines

Consensual writing guidelines are extremely effective in single sourcing projects. Because re-usable content is a fundamental project requirement, the entire team is highly motivated to follow shared guidelines for shared information. Everyone wants their content modules to mesh, not clash. Once the team is clear on the necessity of modular writing, establishing consensual guidelines becomes easy. Each writer brings ideas to the table, where they are evaluated by the rest of the team. Ideas that work for the entire team become guidelines. Because the guidelines come from the team, they are enforced by the team.

Base guidelines on success

Always base your writing guidelines on what has actually worked in real projects. The guidelines should explain, point for point, strategies that worked well in previous projects. Make sure to include examples. If your team does not have any previous experience with single sourcing projects, re-use the usability guidelines you followed in previous online help projects as a basis for your modular writing guidelines.

Update guidelines regularly

Set up your writing guidelines as modular documents that can be updated easily. As team members learn lessons about information re-use in live single sourcing projects, incorporate these lessons into your guidelines. Once your guidelines stabilize to the point where they are no longer controversial, set up a quarterly review process. Ask team members to evaluate the guidelines to make sure they are as current as they appear.

TIP In establishing consensual writing guidelines, it is helpful to have a "neutral" third party, such as an editor, compile the guidelines into a document.

CHAPTER 2

Building documents

At the beginning of a project, the single sourcing process does not look all that different from the traditional documentation process. Before you begin to develop information, you identify your users, the types of information they need, and the types of documents that best communicate that information. Once you begin to develop information, however, the difference between single sourcing projects and traditional projects becomes dramatic. In traditional projects, you develop one type of document at a time. In single sourcing, you develop modular content once, then assemble it into different documents for different audiences and purposes.

Single sourcing is much more than mechanical document assembly. What really drives the single sourcing process is modular writing. Modular writing is a cognitive process. You evaluate content, break it into the smallest possible modules, label the modules by content type, configure the modules into meaningful hierarchies, and link the hierarchies to related hierarchies.

This chapter explains in 10 steps how to transform linear documents into modular documents. The steps are cross-referenced to detailed guidelines and examples in Chapter 3, "Structuring content," and Chapter 4, "Configuring language."

In this chapter

This chapter is designed to help you get started single sourcing in a legacy environment comprised of linear documentation. This chapter assumes you have inherited a linear document that you want to transform into a modular document for a single sourcing project. In effect, it shows you how to bring your linear document up to code for single sourcing.

Because this chapter assumes you are starting with a legacy document, it begins the 10-step process by identifying the modules in your existing document (Step 1). From there, you label the modules (Step 2), then organize them into sections by module type (Step 3). After you organize your single-source document into modular sections, you build (Step 4) and edit (Step 5) each module. Only after developing a clean single-source "database" do you organize (Step 6), cross-reference (Step 7), and convert your input document into different output formats (Step 8). You then test each generated document (Step 9). Finally, based on your experiences, you develop guidelines for your next project (Step 10).

The steps in this chapter constitute a generic process you can modify, not a specific procedure you must follow. Although this process has proven itself in successful single sourcing projects, it is by no means the only approach to single sourcing that works. Modify the content and order of the steps to meet the specific requirements of your company, projects, and users.

TIP If you are building a single-source document from scratch, you might want to change the order of the steps in this chapter. For example, you might want to build modules first, then label them, and then organize them.

This chapter focuses on a typical single sourcing environment in which printed manuals are authored with a desktop publishing application, then converted to online help formats with a document conversion tool. If you use different tools, adapt the steps in this chapter accordingly.

For more about development tools, see "Development tools" on page 188.

This chapter contains the following steps:

Step 1. Identifying modules 26

Identify the primary and secondary modules that make up your existing linear document.

Step 2. Labeling modules 28

Label each module using a syntax that clearly identifies the content of the module, and how that content is presented.

Step 3. Organizing modules 30

In your single-source document, integrate secondary modules into primary modules. Then segregate primary modules by type.

Step 4. Building modules 32

Build each module in your single-source document. Structure modules consistently. Follow guidelines for each module type.

Step 5. Editing modules 34

Edit the modules to ensure that your single source is clean. Use clear, concise, and consistent language that works in any context.

Step 6. Organizing documents 36

Organize primary modules into document hierarchies. Then flatten the hierarchies to bring information to the surface.

Step 7. Cross-referencing documents 38

Cross-reference your document by building a table of contents, section contents, inline cross-references, and an index.

Step 8. Converting documents 40

Map source templates to target templates, then generate new document formats automatically.

Step 9. Testing documents 42

Test the usability of each print or online document you generate. Make sure to test documents, not testers or development methods.

Step 10. Developing guidelines 44

Develop consensual writing guidelines. Base guidelines on solutions uncovered during development, editing, and testing.

Step 1 Identifying modules

Before converting an existing linear document into a modular document, you need to identify the parts, or modules, that make up the document.

Identifying primary modules

Most documents are made up of primary modules, such as the following:

- **Definition lists**
  Definition lists describe product components and technologies.
  For details, see "Definition lists" on page 63.
- **Glossaries**
  Glossaries are definitions lists that explain technical terms used in document modules.
  For details, see "Glossaries" on page 81.
- **Procedures**
  Procedures explain how to perform sequential tasks, step by step.
  For details, see "Procedures" on page 118.
- **Processes**
  Processes describe sequential tasks performed by people or things.
  For details, see "Processes" on page 124.
- **Topics**
  Topics answer who, what, when, where, or why with argument, description, exposition, or narration.
  For details, see "Topics" on page 140.
- **Troubleshooting scenarios**
  Troubleshooting scenarios explain problems (topics) and their solutions (procedures).
  For details, see "Troubleshooting scenarios" on page 144.

Identify all of the primary modules in your document.

TIP You eventually organize primary modules into sections of single-source documents. For details, see Step 3, "Organizing modules," on page 30.

Identifying secondary modules

Most primary modules contain secondary modules, such as the following:

- **Examples**

  Examples are words or phrases that illustrate texts. Short examples appear within texts. Long examples stand alone.

  For details, see "Examples" on page 67.

- **Figures**

  Figures are images or charts that illustrate texts. Images and charts are often developed externally, then referenced in documents.

  For details, see "Figures" on page 74.

- **Itemized lists**

  Itemized lists display serial items vertically, for easy scanning. The lists can be simple or complex, and can include headings and annotations.

  For details, see "Itemized lists" on page 98.

- **Notes**

  Notes are small text blocks that supplement other modules. These text blocks can contain positive or negative messages.

  For details, see "Notes" on page 106.

- **Tables**

  Tables are collections of columns and rows used to compare related information in a small visual space.

  For details, see "Tables" on page 129.

Identify all of the secondary modules in your document.

TIP Although you eventually integrate secondary modules into primary modules, you can build secondary modules separately. For example, to build figures, you can use different applications than those used to build primary modules. Also, it is not uncommon for graphic designers to build images and charts that are re-used in different documents.

In most documents, textual modules deliver the primary messages. Visual modules illustrate or refine those primary messages. There are exceptions to this rule. In picture-driven instructions, like those pioneered by the U.S. Department of Defense, images and charts deliver the primary message, which is supplemented by short texts.

Step 2 Labeling modules

Label each module in your document using a syntax that clearly identifies the content of the module, and how that content is presented.

Labeling primary modules

Build headings for primary modules, such as the following:

- **Definition lists**
  When labeling definition lists, use a consistent heading syntax that distinguishes different types of definitions.
  For details, see "Labeling definition lists" on page 87.
- **Glossaries**
  When labeling glossary divisions, separate alphabetic and numeric characters. Never group characters. Ignore missing characters.
  For details, see "Labeling glossaries" on page 89.
- **Procedures**
  When labeling procedures, answer the question "How?" with a verb. Distinguish between superprocedures and other procedures.
  For details, see "Labeling procedures" on page 90.
- **Processes**
  When labeling processes, answer the question "How?" with a verb. Use gerunds rather than action verbs.
  For details, see "Labeling processes" on page 90.
- **Topics**
  When labeling topics, answer specific questions. Use a consistent syntax that distinguishes different types of topics.
  For details, see "Labeling topics" on page 91.
- **Troubleshooting scenarios**
  When labeling troubleshooting scenarios, describe problems, not their solutions. Describe problems from the user perspective.
  For details, see "Labeling troubleshooting scenarios" on page 91.

TIP Build primary module headings that make sense in *any* table of contents.

Labeling secondary modules

Use captions, headings, or icons to label secondary modules, such as the following:

- **Examples**

  Most examples do not have labels. If an example itself constitutes a stand-alone section or subsection, add a descriptive heading.

  For details, see "Labeling examples" on page 88.

- **Figures**

  When labeling figures, use captions that describe their content from the system or user perspective.

  For details, see "Captions" on page 51.

- **Itemized lists**

  When labeling itemized lists, use a consistent and predictable heading syntax that distinguishes different types of information.

  For details, see "Labeling itemized lists" on page 89.

- **Notes**

  When labeling notes, use words that distinguish between positive advice (notes and tips) and negative advice (cautions and warnings).

  For details, see "Notes" on page 106.

- **Tables**

  When labeling tables, use captions that describe their content from the system perspective only.

  For details, see "Captions" on page 51.

When labeling secondary modules, use a syntax that clearly identifies module content and type.

TIP Although it may seem strange to label secondary modules (for example, itemized lists) that are integrated into primary modules with their own headings, accurate labels make it easier for you to organize the modules. For details, see Step 3, "Organizing modules," on page 30.

Step 3 Organizing modules

In your single-source document, integrate secondary modules into primary modules. Then segregate primary modules into distinct sections.

Integrating secondary modules

Integrate secondary modules into primary modules, as follows:

- **Examples**

  You can add examples to definition lists, itemized lists, procedures, processes, topics, and troubleshooting scenarios.

  For details, see "Integrating examples" on page 69.

- **Figures**

  You can add figures to topics. If possible, do not add figures to other types of primary modules.

  For details, see "Integrating figures" on page 77.

- **Itemized lists**

  You can add itemized lists to definition lists, procedures, processes, section contents, topics, and troubleshooting scenarios.

  For details, see "Integrating itemized lists" on page 102.

- **Notes**

  You can add notes, tips, cautions, and warnings to definition lists, procedures, processes, topics, and troubleshooting scenarios.

  For details, see "Integrating notes" on page 108.

- **Tables**

  You can add tables to topics. If possible, do not add tables to other types of primary modules.

  For details, see "Integrating tables into topics" on page 137.

TIP Secondary modules do not stand alone. If you integrate a secondary module into another secondary module, you must still integrate the "parent" secondary module into a primary module.

Segregating primary modules

Segregate primary modules into distinct sections, as follows:

- **Definition lists**

  Build a section for all definition lists. Definition lists explain product components and technologies (for example, what menu items do).

- **Glossary**

  Build a glossary for all definitions of technical terms used in your single-source document (for example, terms used in topics).

- **Procedures**

  Build a section for all procedures. Procedures explain how users perform actions (for example, how to print a document).

- **Processes**

  Build a section for all processes. Processes explain how products perform specific actions (for example, converting XML content modules into HTML documents), or how users perform more general actions (for example, diagnosing problems).

- **Topics**

  Build a section for all topics. Topics describe who, what, when, where, or why (for example, the purpose of a product).

- **Troubleshooting scenarios**

  Build a section for all troubleshooting scenarios. Troubleshooting scenarios describe problems and explain how to solve them (for example, what to do if an application does not start).

In print-based single-source documents, you can organize procedures, processes, and topics into separate chapters. Likewise, you can group definition lists and troubleshooting scenarios into appendices.

TIP By segregating primary modules into different sections of your single-source document, you build a "database" you can re-use when you organize documents. To make it easier to find information in your single-source document, sort the modules in each chapter or appendix alphabetically. If your headings clearly indicate module content and type, your alphabetized single-source document may actually suggest ways to organize documents. For related steps, see Step 6, "Organizing documents," on page 36.

Step 4 Building modules

Build each module in your single-source document. Structure modules consistently, following guidelines for each module type.

Building primary modules

Build primary modules, such as the following:

- **Definition lists**
  When building definitions, format them as definitions lists, not tables. Wherever possible, use parallel construction.
  For details, see "Definition lists" on page 63.
- **Glossaries**
  When building glossary entries, use abbreviations rather than their spellouts. Never use a term to define itself.
  For details, see "Glossaries" on page 81.
- **Procedures**
  When building procedures, introduce them with standard wording. Structure procedure steps to emphasize user actions and options.
  For details, see "Procedures" on page 118.
- **Processes**
  When building processes, format steps as ordered list items.
  For details, see "Processes" on page 124.
- **Topics**
  When building topics, order content by level of importance and detail. Use active voice, second person, and present tense.
  For details, see "Topics" on page 140.
- **Troubleshooting scenarios**
  When building troubleshooting scenarios, separate problems and solutions. Use topics for problems. Use procedures for solutions.
  For details, see "Troubleshooting scenarios" on page 144.

TIP If you are converting linear documents to modular documents, you *rebuild* (or "modularize") content rather than *build* modular content from scratch.

Building secondary modules

Build secondary modules, such as the following:

- **Examples**

  Format simple examples with parentheses. Format complex examples with paragraphs. Do not include examples within examples.

  For details, see "Examples" on page 67.

- **Figures**

  If you are building print and online documents from a single source, optimize images for print, then convert them to online formats. To maximize re-usability, embed images in documents by reference.

  For details, see "Figures" on page 74.

- **Itemized lists**

  To maximize usability and re-usability, use parallel construction for all items within an itemized list.

  For details, see "Itemized lists" on page 98.

- **Notes**

  Make sure that notes and tips offer positive advice. Make sure that cautions and warnings offer negative advice.

  For details, see "Notes" on page 106.

- **Tables**

  Avoid tables where you could use lists. When introducing tables, use standard wording and electronic cross-references. To eliminate redundant table cells, combine redundant rows and columns.

  For details, see "Tables" on page 129.

TIP When building modules, maintain a consistent point of view by using parallel construction. For details, see "Parallel construction" on page 158.

Step 5 Editing modules

Before organizing input modules into an output document, edit the input modules to make sure your single-source "database" works in any context. This up-front editing eliminates redundant edits of assembled content.

Verifying module labels

Make sure headings and captions are written as follows:

- **Captions**
  Describe figures from the system perspective or the user perspective. Describe tables from the system perspective only.
  For details, see "Captions" on page 51.
- **Headings**
  Make sure headings answer specific questions as directly as possible. Make sure headings make sense when taken out of context.
  For details, see "Headings" on page 85.

Verifying module organization

Make sure primary and secondary modules are organized properly:

- **Primary modules**
  Make sure each type of primary module is organized into a distinct section of your single-source document.
  For guidelines, see "Segregating primary modules" on page 31.
- **Secondary modules**
  Make sure each secondary module is properly integrated into a primary module.
  For guidelines, see "Integrating secondary modules" on page 30.

TIP Module edits do not replace document edits. Once you assemble edited input modules into output documents, edit each output document.
For details, see Step 9, "Testing documents," on page 42.

Verifying module language

Make sure the language in your modules is effective in any context:

- **Abbreviations**

  Spell out abbreviations the first time they are mentioned in a module. Use common abbreviations, but not uncommon abbreviations.

  For details, see "Abbreviations" on page 151.

- **Capitalization**

  Capitalize document titles, captions, commands, definition lists, filenames, glossaries, and itemized lists consistently.

  For details, see "Capitalization" on page 154.

- **Parallel construction**

  Use parallel construction to make your modules easier to scan.

  For details, see "Parallel construction" on page 158.

- **Person**

  Use second person singular to speak directly to users.

  For details, see "Person" on page 167.

- **Sentence construction**

  Shorten and simplify sentences. If a sentence contains many serial items, format the serial items as an itemized list.

  For details, see "Sentence construction" on page 171.

- **Tense**

  Use present tense to maintain user-centered time in modules.

  For details, see "Tense" on page 173.

- **Voice**

  Use active voice to increase the clarity of modules.

  For details, see "Voice" on page 177.

TIP For a structured approach to editing, refer to *The Levels of Edit* by the late Robert Van Buren and Mary Fran Buhler. Originally published in 1976 by the Jet Propulsion Laboratory (JPL), this booklet is now available from the Society for Technical Communication (STC).

Step 6 Organizing documents

Organize primary modules into document sections and subsections for a specific audience, purpose, and format. Then flatten the structure of your document to bring information to the surface and increase usability.

Building document hierarchies

To build document sections and subsections, organize primary modules by any of the following methods:

- **Alphabet**

  Organize modules alphabetically by heading. For example, order procedures and definition lists alphabetically for easy reference.

- **Audience**

  Organize modules by type of user. For example, separate sections for novices and experts.

- **Detail**

  Organize modules by level of detail. For example, introduce general product features before explaining specific product functions.

- **Importance**

  Organize modules by level of importance. For example, list installation and de-installation procedures before other procedures.

- **Location**

  Organize modules by the physical location of the objects they describe. For example, introduce the parts of an application as they appear in the user interface, from left to right, and from top to bottom.

- **Sequence**

  Organize modules in sequential order. For example, include installation procedures before configuration procedures.

- **Type**

  Organize modules by type. Segregate primary modules into distinct sections. For example, separate topics and procedures.

For guidelines, see "Organizing output documents" on page 112.

Refining document hierarchies

Complex hierarchies can help you, as a writer, to understand complex products. But complex hierarchies can actually hinder users from using those same products. This usability problem is especially acute in online documents. Online documents do not provide users with enough visual space to see complex hierarchies.

To refine document sections and subsections, flatten hierarchies using one of the following methods:

- **Promoting sections**

  An easy way to simplify hierarchies is to promote sections. For example, to simplify a chapter with many different heading levels, convert each section (first-level heading) to a chapter. In effect, you convert one complex hierarchy into many simple hierarchies.

  For details, see "Promoting sections" on page 116.

- **Promoting subsections**

  A not-so-easy way to simplify hierarchies is to promote subsections. For example, if you want to simplify a chapter with many different heading levels, but cannot convert sections to chapters, convert each subsection (second-level heading) to a section (first-level heading).

  For details, see "Promoting subsections" on page 117.

TIP Before you promote subsections, make sure to build section contents that link sections to subsections. For details, see "Section contents" on page 126.

Test hierarchies in online formats before testing them in print formats. For more about testing, see Step 9, "Testing documents," on page 42.

For a comparison of usability challenges in print and online formats, see "Improving document usability" on page 9.

Step 7 Cross-referencing documents

Cross-reference your document by building (generating) a table of contents, section contents, inline cross-references, and an index. Set up preformatted styles with electronic hyperlinks that can be re-used (regenerated) in print and online documents.

TIP For related guidelines, see "Automating cross-references" on page 54.

Building a table of contents

Tables of contents logically link the primary modules that make up a document. No matter what their format, all documents require tables of contents that show the document structure. Build a table of contents that provides users with a summary view or a detailed view of your document. Include chapters and appendices, as well as their subsections. Reference the glossary and index as such.

For details, see "Tables of contents" on page 138.

TIP In printed documents, tables of contents are grouped together with title pages, copyright pages, prefaces, and introductions as "front matter." To find out how to re-use front matter, see "Front matter" on page 79.

Building section contents

Section contents are lists that introduce subsections. These lists are especially important in online documents, where they link document sections to their subsections. In each section that contains a subsection, build an annotated table of contents. This table can be a list that contains the subsection title, a brief description of the subsection, and a page number or a hyperlink to the referenced subsection.

For details, see "Section contents" on page 126.

TIP Section contents are especially useful in online documents, which often split sections and subsections into separate files. In these online documents, section contents hyperlink files to subsections.

Building inline cross-references

Inline cross-references link related modules in a document to each other. In printed documents, cross-references normally include page numbers. In online documents, they do not. Add inline cross-references from modules to related modules (for example, from a procedure to a definition list, and vice versa). Set up preformatted styles and electronic hyperlinks you can re-use in different print and online formats.

For details, see "Cross-references" on page 54.

TIP Cross-references enable you to eliminate redundant information, and to separate primary modules in your documents. For example, if you define command options once in an appendix, you can build cross-references to these definitions from topics and procedures.

For related guidelines, see the following steps:

- Step 3, "Organizing modules," on page 30
- Step 6, "Organizing documents," on page 36

Building an index

Indexes are alphabetized cross-reference lists that enable users to locate information quickly. These lists show relationships between topics, procedures, definitions, and so on. Build a comprehensive index for your document. Build no more than three levels of index entries. If possible, use an indexing tool that enables you to tag modules and generate indexes dynamically, then convert the generated indexes into different formats automatically.

For details, see "Indexes" on page 92.

Step 8 Converting documents

Document conversion is the mechanical process of transforming documents from one format to another format. Before you can convert your document, you need to build templates for different formats, then map the templates to each other. Once you build and map templates, you can generate new documents in different formats automatically.

Types of document formats

You can convert your document into many different document formats, such as the following:

- **Printed manuals**
  Documents designed to be printed (for example, user guides, reference manuals, and so on). These documents can be bound books or print-ready electronic books.
- **Online help**
  Documents designed in any number of different online help formats (for example, Microsoft HTML Help, Microsoft WinHelp, Oracle Help for Java, Sun JavaHelp, and so on). These documents can be context-sensitive help or online documentation.
- **Training materials**
  Documents designed for onsite or offsite training sessions (for example, slide shows, workbooks, and so on). These documents can appear in different print and online formats.
- **Websites**
  HTML-based documents displayed as webpages on the Internet. These documents should not be confused with HTML-based online documentation, which consists of webpages viewed locally by users.

These document formats do not change modular content. They simply determine in which order and in what format the modules are presented.

TIP Do not confuse conversion with assembly, re-assembly, or repurposing. Think of assembly as organization. Think of re-assembly as re-organization. And think of repurposing as "re-assembly lite." For more about repurposing and re-assembly, see "Types of single sourcing" on page 15.

Converting a document to another format

NOTE This process assumes you are using document-driven tools for single sourcing. If you are using database-driven tools, you may follow a different process. For an overview of tools, see "Development tools" on page 188.

To convert a document from one format to another, you follow a process such as the following:

1. **Build a source template.**

   Build a template for your single-source document using a document development tool (for example, Adobe FrameMaker+SGML). Make sure to use special features (for example, conditional text) to prepare the content of the single-source document for conversion to another format (for example, from a printed manual to an online help system).

   For related guidelines, see "About authoring tools" on page 189.

2. **Build a target template.**

   Build a template for the second document you want to generate from the single-source document:

   a. Build a prototype template in the native environment of the second document (for example, Microsoft HTML Help).

   b. Build a conversion template using a document conversion tool (for example, Quadralay WebWorks Publisher).

   For related guidelines, see "About conversion tools" on page 190.

3. **Map the source template to the target template.**

   Map elements in the source template (for example, SGML elements) to elements in the target template (for example, HTML tags).

4. **Generate the target document.**

   Generate the target document (for example, HTML Help) using the document conversion tool (for example, WebWorks Publisher).

TIP To streamline your conversion process, establish an information architect to select, customize, and maintain single sourcing tools for your team. For details, see "Successful single sourcing" on page 19.

Step 9 Testing documents

Test each document you assemble. In document edits, pay close attention to document organization and cross-references. In usability tests, focus on results, not development methods. Correct all problems you find.

TIP If you do not have the resources to test the print and online formats of a given document, test the online format only. Online formats present a worst-case usability scenario. What works online works better on paper.

Editing documents

When editing documents, verify their organization and cross-references. Correct all document-specific problems.

Verifying document organization

Make sure the document is organized in a way that highlights your message, and that is easy to understand.

For guidelines, see the following sections:

- "Organizing output documents" on page 112
- "Flattening hierarchies" on page 115

Verifying document cross-references

Make sure the document is fully cross-referenced in a way that makes it easy to use. Replace redundant information with cross-references.

For guidelines, see the following sections:

- "Tables of contents" on page 138
- "Section contents" on page 126
- "Cross-references" on page 54
- "Indexes" on page 92

TIP If you need to edit modular content, make the changes in your single-source document, not in your output documents. For more information about module editing, see Step 5, "Editing modules," on page 34.

Testing document usability

Test document accessibility and usefulness internally and externally.

Types of testers

Depending on your resources, test documents internally or externally:

- **Internal testers**

  Ask internal users (for example, coworkers) to test your document. Choose testers who are not part of the product team.

- **External testers**

  Ask external users (for example, product beta testers) to test your documents. Choose testers who are familiar with the product.

If possible, test documents internally, then externally.

Types of tests

Test the usability of each output document:

- **Accessibility**

  Information is useless if users cannot find it. Test whether users can find the information they need. Ask them to use your document to solve specific problems. Then watch closely to see whether they can find the information they need.

- **Usefulness**

  Information is useless if it does not answer user questions. Once users find information, watch closely to see if they can use the information to solve the problems you have given them.

Make sure to integrate your usability results into your writing guidelines. To get started, see Step 10, "Developing guidelines," on page 44.

TIP In usability tests, pay attention to when and where users stop. If users stop, it is an indication that there is a problem with the documentation.

No matter how you conduct your usability tests, it is critical that users know that the document, not they, are being tested. One simple way to "turn the tables" on your document is to ask two users to test the document together. Typically, they "gang up" on the document, expressing their pleasure and displeasure out loud. You hear what they really think in no uncertain terms.

Step 10 Developing guidelines

Develop consensual writing guidelines, based on problems and solutions uncovered during the development, editing, and usability test cycles. You can develop writing guidelines from the top down, or from the bottom up. Bottom-up guidelines are more appropriate to single sourcing projects because they are based exclusively on what works in actual projects, not what should work in theoretical projects. Consensual writing guidelines reinforce team synergy in single sourcing projects.

Developing top-down guidelines

You can develop documentation standards for an entire corporation. Corporate style guides are often developed by committee. The committee develops guidelines based on commonly accepted academic standards (for example, *The Chicago Manual of Style*) and commonly accepted industry standards (for example, the corporate style guides published by large software manufacturers). The committee distributes draft chapters for review to a subcommittee, composed of publications managers. If a majority of the publication managers approve of the draft guidelines, they are published company-wide, and become law for writers and editors.

Developing bottom-up guidelines

You can develop writing guidelines for a specific project. Project style guides are often developed by an editor already assigned to the project. The editor develops guidelines based on current edits of project modules or documents. The guidelines are a summary of successive agreements between the editor and individual writers. As the project progresses, the guidelines grow. Once the project is completed, the editor and writers hold a meeting to discuss what worked and what did not work in the project. The editor then compiles the "success stories" as guidelines into a formal project style guide to be re-used for future projects.

TIP The best writing guidelines are flexible structures based on actual success. Such guidelines are updated regularly, to keep pace with project realities.

Developing consensual guidelines

Consensual writing guidelines are extremely effective in single sourcing projects. Because re-usable content is a fundamental project requirement, the entire team is highly motivated to follow shared guidelines for shared information. Everyone wants their content modules to mesh, not clash.

Developing consensual writing guidelines is a cyclical process with distinct phases, such as the following:

1. **Propose**

   Propose writing guidelines, based on edits, usability test results, and ideas from individual writers. Make sure to include negative and positive examples.

2. **Approve**

   Ask team members to vote on each proposed guideline. Give team members a reasonable time (for example, two weeks) to review the guidelines. Make it clear that no response counts as approval.

3. **Compile**

   Compile a project style guide that includes only those guidelines approved by every member of the team. Unanimity ensures that every team member follows the guidelines in the future.

4. **Distribute**

   Distribute the approved style guide to every member of the team. Make sure to send copies to other publications groups, as well as to the coordinator of your corporate style guide.

5. **Enforce**

   In future projects, empower editors to enforce the guidelines. Encourage writers and editors to propose new guidelines.

Repeat this process for every single sourcing project.

TIP Base your writing guidelines on what has actually worked in real projects. Explain, point for point, strategies that worked well in previous projects. Make sure to include examples.

For sample writing guidelines you can re-use in your own projects, see Chapter 3, "Structuring content," and Chapter 4, "Configuring language."

CHAPTER 3

Structuring content

Modular writing drives the single sourcing process. If you develop modular content that makes sense in any context, you can re-use it in different formats for different audiences and purposes.

This chapter explains in detail and by example how to structure modular content. It explains how to develop primary modules (for example, procedures and topics) and secondary modules (for example, figures and tables). It explains how to organize primary and secondary modules in your single-source document, then re-organize them as output documents in print and online formats. And it explains how to build "cognitive bridges" between assembled modules (for example, tables of contents, section contents, cross-references, and indexes).

For easy reference, the guidelines presented in this chapter are organized alphabetically. Each guideline is illustrated with negative and positive examples, and is cross-referenced to related guidelines. Although each guideline has proven itself in successful single sourcing projects, not every guideline is suitable for every project. Choose the guidelines you need. Then modify them to fit your corporate and project requirements.

In this chapter

This chapter contains the following guidelines:

Captions 51

Captions are titles to figures and tables. Include caption numbers in print. Do not include caption numbers online. Describe figures from the system perspective or the user perspective. Describe tables from the system perspective only.

Cross-references 54

Cross-references link the modules that make up a document. In printed documents, include page numbers in cross-references. In online documents, hyperlink cross-references. To develop cross-references that are effective in print and online documents, set up preformatted styles and electronic hyperlinks.

Definition lists 63

Definition lists define product components and technologies. To maximize re-usability, format terms and definitions as definition lists, not tables. When introducing definition lists, use standard wording. Use terms exactly as they appear in the product you are documenting. Use parallel construction in definitions.

Examples 67

Examples are words and phrases that illustrate texts. Unlike simple examples, complex examples are too large to appear within sentences. Separate simple examples with parentheses. Separate complex examples with paragraph breaks. Do not nest examples.

Figures 74

Figures are images or charts. When building print and online documents from a single source, optimize images for print, then convert them to online formats. To maximize re-usability, embed images into documents by reference.

Front matter 79

Front matter is material that appears at the front of a book, before content begins. Although front matter has its origins in print media, it has parallels in online media. You can re-assemble front matter from printed formats in online formats, and vice versa.

Glossaries **81**

Glossaries explain the technical terminology in a document. In terms, use abbreviations rather than their spellouts. Begin definitions with nouns or verbs. Develop master glossaries to increase consistency and eliminate redundancy.

Headings **85**

Headings are titles to document sections. In headings, answer specific questions as directly as possible. Use a heading syntax that indicates the module type and content. Make sure headings are understandable when taken out of context.

Indexes **92**

Indexes are alphabetized cross-reference lists that help users locate information quickly. Wherever possible, nest index entries. Format index entries for print or online formats. To help users find information in document sets, build master indexes.

Itemized lists **98**

Itemized lists display serial items vertically for easy scanning. The lists can include headings, annotations, and other lists. Use parallel construction for all items within an itemized list. Convert long sentences with serial words or phrases to lists.

Notes **106**

Notes are small text blocks that supplement other modules. These text blocks contain positive or negative advice. For positive advice, use a note or tip. For negative advice, use a caution or warning.

Organization **111**

Document organization drives document assembly. Organize content modules into documents that target specific audiences, purposes, and formats. Segregate primary modules, integrate secondary modules, and flatten hierarchies.

Procedures **118**

Procedures are step-by-step instructions that explain how to perform actions. There are four types of procedures: single-step procedures, multiple-step procedures, superprocedures, and subprocedures. Emphasize user actions and options.

Processes 124

Processes answer one question: "How?" Because the answers are sequential, they are often formatted as ordered lists. Like topics, processes are descriptive, not imperative. They explain what someone or something does, not what users should do.

Section contents 126

Section contents are lists that introduce subsections. The most common format is an itemized list. In online documents, section contents supplement the main navigation system by hyperlinking document sections to their subsections. For complex sections, add annotations. If possible, include electronic cross-references.

Tables 129

Tables are collections of columns and rows that compare related information in a small visual space. Avoid tables where you could use lists. When introducing tables, use standard wording and electronic cross-references. To eliminate redundant table cells, combine redundant rows and columns.

Tables of contents 138

Tables of contents list the modules that make up a document. List all prefaces and introductions, but not their sections. List chapter and appendix sections for clarity (sections only) or completeness (sections and subsections). List glossaries and indexes as such.

Topics 140

Topics are texts that describe who, what, when, where, or why. To make your answers clear, order sentences and paragraphs by level of importance and detail. To speak directly to users, write topics in active voice, second person, and present tense.

Troubleshooting scenarios 144

Troubleshooting scenarios are a hybrid of topics and procedures. Introduce scenarios briefly, then explain problems and solutions in detail. Format problems as topics. Format solutions as procedures. Group closely related problems and solutions.

Captions

Captions are titles to figures and tables. Include caption numbers in print documents, but not in online documents. In figure captions, describe content from the system perspective or from the user perspective. In table captions, describe content from the system perspective only.

TIP For related guidelines, see the following sections:

- "Capitalizing captions" on page 155
- "Figures" on page 74
- "Parallel captions" on page 158
- "Tables" on page 129

Types of captions

Captions are titles to figures or tables:

- **Figures**

  Captions describe charts (for example, bar charts, pie charts, or flow charts) or images (for example, photographs or screenshots). Figure captions should describe content from the system (object) perspective or from the user (action) perspective.

- **Tables**

  Captions describe textual information. In print documents, table captions normally include "*(continued)*" on every page after the first page. Table captions should describe content from the system (object) perspective only.

TIP Most authoring tools enable you to build lists of figures and tables in much the same way that you build tables of contents. Although most technical publications do not include lists of figures or tables, you can build both types of lists temporarily to make sure your captions are parallel.

For related guidelines, see the following sections:

- "About authoring tools" on page 189
- "Parallel construction" on page 158

Types of numbering

In print documents, captions are numbered sequentially for easy reference. In online documents, captions are rarely numbered.

When numbering captions, you can use absolute or relative numbering:

- **Absolute (document)**

  Absolute caption numbers reflect the absolute sequence of the figures and tables in a document (for example, "Figure 1").

- **Relative (chapter or appendix)**

  Relative caption numbers reflect the relative sequence of figures and tables within a chapter or appendix (for example, "Figure 1-1").

Whichever scheme you choose, keep figure and table numbers separate.

TIP Most authoring and conversion tools enable you to number figure and table captions automatically. In addition, they enable you to define different caption styles (for example, with absolute or relative numbering).

Numbering print captions

In print documents, include absolute or relative caption numbers. These numbers enable you to cross-reference figures and tables by number rather than title. As a rule, it is easier for users to find numbers than titles.

Change	To
Once you have entered your new password, you receive a system message confirming your password change, as shown in *the "System message confirming password change" figure.* Figure: System message confirming password change	Once you have entered your new password, you receive a system message confirming your password change, as shown in *Figure 1.* Figure 1 System message confirming password change
The table on page 357 shows the options evaluated by AcmePro in the startup scripts. Table: Startup script options	*Table A-1 on page 357* shows the options evaluated by AcmePro in the startup scripts. Table A-1 Startup script options

Numbering online captions

In online documents, do not include caption numbers. Normally, online documents are not viewed sequentially.

Change	To
Table 13-1 on page 124 shows the encapsulated logfiles and associated templates used on Sun Solaris managed nodes.	Table: Encapsulated logfiles and templates shows the encapsulated logfiles and associated templates used on Sun Solaris managed nodes.
Table 13-1 Encapsulated logfiles and templates	Table: Encapsulated logfiles and templates

Wording figure titles

When wording figure titles, use consistent syntax. Describe figures from the system perspective (for example, begin captions with nouns) or from the user perspective (for example, begin captions with verbs).

System perspective	User perspective
Figure 1-2 Print dialog box	Figure 1-2 Printing a document

Wording table titles

When wording table titles, use consistent syntax. Describe tables from the system perspective only (for example, begin captions with nouns).

Change	To
Table A-1 Viewing object tree items in the shortcut bar	Table A-1 Object tree items in the shortcut bar

Cross-references

Cross-references link the modules that make up a document. In printed documents, cross-references almost always include page numbers. In online documents, cross-references never include page numbers. To build cross-references you can re-use in print and online formats, set up preformatted styles and electronic hyperlinks.

Types of cross-reference formats

Essentially, there are two different formats for cross-references:

- **Print**

  Cross-references in printed documents include page numbers. Users find the referenced information by turning to the pages indicated.

- **Online**

  Cross-references in online documents are hyperlinked. Users access the referenced information by clicking the hyperlinks.

Automating cross-references

In linear documentation, you develop print and online cross-reference formats separately. As a result, it is impossible to re-use cross-references developed in print documents for online documents, and vice versa.

In single sourcing, you can set up preformatted styles with electronic hyperlinks. Preformatted styles ensure consistent syntax for your cross-references. In effect, you can build writing guidelines into the software you use to develop documents. Electronic hyperlinks enable you to convert cross-references to different print and online formats. Electronic cross-references also automate updates to tables of contents, section contents, inline cross-references, and indexes.

TIP Most authoring tools (for example, Adobe FrameMaker) enable you to automate cross-references. You develop styles for different types of cross-references, then update them automatically. These different cross-reference styles are based on hyperlinks. You can convert the styles to online formats with conversion tools (for example, Quadralay WebWorks Publisher). For more about development tools, see "Development tools" on page 188.

Cross-referencing definition lists

In definition lists, cross-reference terms to related modules, rather than adding the related modules to the definition lists.

Change

About the File menu

The File menu in the menu bar has the following options:

`Save Console Session Settings`

> Stores customizations you made in the Preferences dialog box.
>
> The Preferences dialog box contains the following options:
>
> `Refresh Interval`
>
> > Determines how frequently the GUI is automatically refreshed.
>
> `Message Display`
>
> > Determines how many messages to display in the message browser.

`Print`

> Prints the selected messages.
>
> To print messages, follow these steps:
>
> 1. In the message browser, select the messages you want to print.
> 2. In the File menu, select `Print`.

To

About the File menu

The File menu in the menu bar has the following options:

`Save Console Session Settings`

> Stores customizations you made in the Preferences dialog box. *For details, see "About the Preferences dialog box" on page 91.*

`Print`

> Prints the selected messages. *For details, see "To print messages" on page 37.*

About the Preferences dialog box

The Preferences dialog box contains the following options:

`Refresh Interval`

> Determines how frequently the GUI is automatically refreshed.

`Message Display`

> Determines how many messages to display in the message browser.

To print messages

To print messages, follow these steps:

1. In the message browser, select the messages you want to print.
2. In the File menu, select `Print`.

TIP For related guidelines, see "Definition lists" on page 63.

Cross-referencing documents

When cross-referencing a document from within itself, use a generic term (for example, "this guide") rather than the title of the document. When cross-referencing related documents, refer to document titles, as section titles can change without notice.

Change	To
The *AcmePro Concepts Guide* helps you better understand and use AcmePro.	This guide helps you better understand and use AcmePro.
For more information about non-standard installations, see Chapter 3, "About non-standard installations," in the *AcmePro Installation Guide.*	For more information about non-standard installations, refer to the *AcmePro Installation Guide.*

TIP To update document titles automatically, use variable text. For details, see "Variables" on page 200.

Cross-referencing figures

References to physical location (above, below, and so on) in a document assume that users read documents sequentially, a very uncommon practice. In printed documents, cross-reference figures by number. In online documents, cross-reference figures by title.

Change	To
The object pane is shown above.	The object pane is shown in Figure 2-3 on page 15.
	The object pane is shown in Figure: Object pane.

TIP For related guidelines, see the following sections:

- "Captions" on page 51
- "Figures" on page 74

Cross-referencing glossaries

In the body of a document, cross-reference terms defined in the glossary. In the glossary itself, build "see" and "see also" references between terms.

Cross-referencing glossary terms in text

In online documents, build cross-references from technical terms used in content modules (for example, procedures and topics) to their definitions in your glossary.

Topic	Glossary
Most hardware and software uses the ASCII character set.	**ASCII** American Standard Code for Information Interchange. Predominant character set encoding of present-day computers....

Building "see" references in glossaries

In glossaries, build "see" references that point from aliases to preferred terms. If there are two or more "see" references, alphabetize the references, and separate them with semicolons (;).

Change	To
browser 1. message browser. Part of the AcmePro user interface that enables you to view messages received by the management server. 2. web browser. Program that enables you to read HTML pages.	**browser** *See* message browser; web browser. **message browser** Part of the AcmePro user interface that enables you to view messages received by the management server. **web browser** Program that enables you to read HTML pages.

Change	To
Cascading Style Sheet Extension to HTML that allows you to define styles for certain elements of an HTML document.	**Cascading Style Sheet** *See* CSS. **CSS** Cascading Style Sheet. Extension to HTML that allows you to define styles for certain elements of an HTML document.

Building "see also" references in glossaries

In glossaries, build "see also" references that point to related terms. If there are two or more "see also" references, alphabetize the references, and separate them with semicolons (;).

Change	To
backup manager Management server that replaces another management server (for example, in case of failure). **management server** Central computer system or intelligent device monitored or controlled by AcmePro. **primary manager** Management server that is currently responsible for agents.	**backup manager** Management server that replaces another management server (for example, in case of failure). *See also* management server; primary manager. **management server** Central computer system or intelligent device monitored or controlled by AcmePro. *See also* backup manager; primary manager. **primary manager** Management server that is currently responsible for agents. *See also* backup manager; management server.

TIP For related guidelines, see "Glossaries" on page 81.

Cross-referencing indexes

In indexes, build "see" and "see also" references between entries.

Building "see" references in indexes

In indexes, build "see" references that point from aliases to preferred terms. Alphabetize the references, and separate them with semicolons (;).

Change	To
SPARCclassic configuring, 398 installing, 397 SPARCstation configuring, 402 installing, 401	SPARCclassic configuring, 398 installing, 397 SPARCstation configuring, 402 installing, 401 Sun Microsystems. *See* SPARCclassic; SPARCstation

Building "see also" references in indexes

In indexes, use "see also" references that point to related terms. Alphabetize the references, and separate them with semicolons (;).

Change	To
configuring managed nodes AIX, 25-26 Linux, 27-28 Solaris, 29-30 installing agent software AIX, 49-50 Linux, 51-52 Solaris, 53-54	configuring managed nodes *See also* installing agent software AIX, 25-26 Linux, 27-28 Solaris, 29-30 installing agent software *See also* configuring managed nodes AIX, 49-50 Linux, 51-52 Solaris, 53-54

TIP For related guidelines, see "Indexes" on page 92.

Cross-referencing procedures

In procedures, cross-reference specific steps, not entire procedures.

Change	To
To upgrade to AcmePro 7.0, follow these steps: 1 Download the configuration. 2 ... *More information about each step is found below.*	To upgrade to AcmePro 7.0, follow these steps: 1 Download the configuration. *For details, see "Download the configuration" on page 146.* 2 ...

TIP For related guidelines, see "Procedures" on page 118.

Cross-referencing tables

References to physical location (above, below, and so on) in a document assume that users read documents sequentially, a very uncommon practice. In printed documents, cross-reference tables by number. In online documents, cross-reference tables by title.

Change	To
For a comparison of managed node libraries for AcmePro 4, 5, and 6, see the table on the previous page.	For a comparison of managed node libraries for AcmePro 4, 5, and 6, see Table 11-6 on page 298.
For a comparison of managed node libraries for AcmePro 4, 5, and 6, see <u>the table above</u>.	For a comparison of managed node libraries for AcmePro 4, 5, and 6, see <u>Table: Managed node libraries</u>.

TIP For related guidelines, see the following sections:

- "Captions" on page 51
- "Tables" on page 129

Cross-referencing topics

In topics, cross-reference related modules, rather than embedding the modules in the topics.

Change

About the message browser

AcmePro provides information about your managed environment through messages displayed in the message browser. A message represents an event that has occurred on a node within the managed environment generated by the agents running on the node.

To provide you with additional information that helps you solve problems, messages are characterized by attributes summarized in the message headline of the message browser.

By default, the message browser headline contains the following elements:

`Time Received`

: Date and time the problem occurred.

`Node`

: Managed node on which the problem occurred.

To

About the message browser

AcmePro provides information about your managed environment through messages displayed in the message browser. A message represents an event that has occurred on a node within the managed environment generated by the agents running on the node.

To provide you with additional information that helps you solve problems, messages are characterized by attributes summarized in the message headline of the message browser. *For a description of default elements in the message headline, see "About the message browser headline" on page 90.*

About the message browser headline

By default, the message browser headline contains the following elements:

`Time Received`

: Date and time the problem occurred.

`Node`

: Managed node on which the problem occurred.

TIP For related guidelines, see "Topics" on page 140.

Cross-referencing troubleshooting scenarios

In troubleshooting modules, cross-reference the entire modules, not their problem descriptions. Cross-reference solution steps, not entire solutions.

Change	To
Cannot start an application If you can no longer start an application on a managed node, adapt the default application startup configuration or use customized startup options. **Problem** An application has been upgraded, and its command path has been changed. *See also "Cannot stop an application" on page 430.*	**Cannot start an application** If you can no longer start an application on a managed node, adapt the default application startup configuration or use customized startup options. *For related scenarios, see "Cannot stop an application" on page 430.* **Problem** An application has been upgraded, and its command path has been changed.
Solution *For a description of the* `ps` *command, see "About the ps command" on p. 256. For a description of the* `kill` *command, see "About the kill command" on p. 257.* Do the following: 1 Use the `ps` command to determine the process ID of the endlessly running action. 2 Use a `kill` command for the specific process ID.	**Solution** Do the following: 1 Use the `ps` command to determine the process ID of the endlessly running action. *For a description of the* `ps` *command, see "About the ps command" on p. 256.* 2 Use a `kill` command for the specific process ID. *For a description of the* `kill` *command, see "About the kill command" on p. 257.*

TIP For related guidelines, see "Troubleshooting scenarios" on page 144.

Definition lists

Definition lists are used to define the hardware or software components that make up products, as well as technologies related to those products. To maximize re-usability, format terms and definitions as definition lists, not tables. When introducing definition lists, use standard wording. Use terms exactly as they appear in the product you are documenting. As a rule, order terms alphabetically. In definitions, use parallel construction.

TIP For related guidelines, see the following sections:

- "Active voice in definition lists" on page 177
- "Capitalizing definition lists" on page 155
- "Cross-referencing definition lists" on page 55
- "Labeling definition lists" on page 87
- "Parallel list items" on page 162

Types of definition lists

Definition lists are used to define product components and related technologies:

- **Components**
  Describe hardware (for example, CPU, monitor, printer, and so on) or software (for example, menu items, commands, variables, and so on). Objects are normally described in stand-alone definition lists.
- **Terminologies**
  Describe proprietary terms (for example, Network Node Manager) or public domain terms (for example, File Transfer Protocol). Technical terms are normally contained in glossaries.

Definition lists group physically or logically related product components. In software documentation, definition lists are used to define related groups of user interface elements. Because these definition lists are often large and complex, separate them from other modules.

TIP The most common kind of definition list is a glossary, which defines all technical terms contained within a given document. For details, see "Glossaries" on page 81.

Formatting definition lists

Format terms and definitions as definition lists, not tables:

- **Lists**

  Definition lists use text formatting to pair terms with their definitions. A typical example of a definition list is a glossary, which looks and feels like a dictionary. Entries are sorted by terms. Terms can be paired with multiple definitions. Because these lists do not tie content to a specific output format, they are easy to re-use in different formats.

- **Tables**

  Definition tables use columns to pair terms with their definitions. Formatting terms and definitions in this way is actually a misuse of tables. Tables tie content to a specific format. As a result, tables are difficult to re-use in different formats.

TIP For related guidelines, see "Avoiding tabular lists" on page 131.

Introducing definition lists

When introducing definition lists, use standard wording. Aim introductions in the direction of the list you are introducing.

Change	To
The File menu stores customizations you made during the current session, and enables you to set preferences for your printer.	The File menu contains the following options:
The following options are contained in the Edit menu:	The Edit menu contains the following options:
The View menu contains:	The View menu contains the following options:

Building list terms

In definition lists, present terms exactly as they appear in the product you are documenting. If components contain subcomponents (for example, if a menu item contains subitems), "nest" the subcomponents under their "parent" components. For easy reference, order terms alphabetically, unless there is a compelling reason not to (for example, logically placing "yes" before "no").

Change	**To**

`Filtering`
Enables you to create and modify filtered message browsers to meet your specific requirements. The `Active Message Browser` option opens the Filter Messages dialog box, where you can create a new active message browser. The `History Message Browser` option opens the Filter Messages dialog box, where you can create a new history message browser.
	`Filtering`
	Enables you to create and modify filtered message browsers to meet your specific requirements:
	`Active Message Browser`
	Opens the Filter Messages dialog box, where you can create a new active message browser.
	`History Message Browser`
	Opens the Filter Messages dialog box, where you can create a new history message browser.

`ActiveX Container`
This option indicates whether or not you want the workspace to be an ActiveX control. If the answer is in the affirmative, the container will become an ActiveX control. If the answer is in the negative, the workspace will not become an ActiveX control.
	`ActiveX Container`
	Indicates whether the workspace is an ActiveX control:
	`Y`
	Yes. Workspace is an ActiveX control.
	`N`
	No. Workspace is *not* an ActiveX control.

Building list definitions

Begin list definitions with nouns or verbs, not articles. Whenever possible, use parallel construction. Never use a term to define itself. Instead, define the term with a synonym, and use the synonym throughout the definition. If definitions contain serial items, convert the items to itemized lists.

Change	**To**
`Print` The Print menu item prints selected messages, all messages in the message browser, details of selected messages, or details of all messages in the message browser.	`Print` Enables you to print the following types of messages from the message browser: • All messages • Selected messages • Details of all messages • Details of selected messages

`Message Key`
Message key that is associated with the message.
`Message No.`
Message number for the message. This message number is useful when you want programming access using open APIs.
	`Message Key`
	Key associated with the message.
	`Message No.`
	Unique identification number associated with the message. This number enables you to program with open APIs.

TIP For related guidelines, see the following sections:

- "Itemized lists" on page 98
- "Parallel list items" on page 162

Examples

Examples are words or phrases that illustrate texts. Simple examples appear within sentences or phrases. Complex examples are too large to appear within sentences. Separate simple examples with parentheses. Separate complex examples with paragraph breaks. Never nest examples.

Formatting simple examples

Place simple examples in text within parentheses. Parentheses provide a clear visual distinction between rules and examples of those rules. If the examples are full sentences, format them as separate sentences.

Change	To
Product names are generally listed according to their respective product number, *for example, AP7000.*	Product names are generally listed according to their respective product number *(for example, AP7000).*
Which items appear in popup menus depends on the object you select in the shortcut bar, *for example, the popup menu items available from a node might be different from the popup menu items available from an URL shortcut.*	Which items appear in popup menus depends on the object you select in the shortcut bar. *For example, the popup menu items available from a node might be different from the popup menu items available from an URL shortcut.*

NOTE At first glance, examples may seem too trivial to mention when discussing single sourcing. In fact, "trivial" examples are precisely the sort of document elements you need to standardize to ensure seamless single sourcing.

If writers in your group write and format examples differently, it will be apparent to users that your documentation has multiple authors. Likewise, if writers use different DocBook elements to format examples (for example, `ComputerOutput`, `Interface`, and `Literal`), converting modules from one format to another is much more difficult.

Formatting complex examples

Format complex examples as paragraphs. Use separate paragraphs for examples you cannot read out loud comfortably.

Change	To
If you use xsl:apply-templates to process elements that are not descendants of the current node, you can create an endless loop. For example, <xsl:template match="test"> <xsl:apply-templates select="."/> </xsl:template> would result in a non-terminating loop.	If you use xsl:apply-templates to process elements that are not descendants of the current node, you can create an endless loop. Example: <xsl:template match="test"> <xsl:apply-templates select="."/> </xsl:template>
For example, you could set up an application with the URL http://www.domain.com/$MSG.TEXT[2].	For example, you could set up an application with the following URL: http://www.domain.com/$MSG.TEXT[2]

Avoiding nested examples

Do not "nest" examples. That is, do not include examples within examples. Although logical, such nesting is difficult for users to follow.

Change	To
For example, when a problem occurs at a managed node (*for example*, an unauthorized user attempts to log in to a managed node), the node registers this problem in one of several possible ways.	*For example*, if a problem occurs on a managed node, the node registers this problem in one of several possible ways.
	For example, if an unauthorized user attempts to log in to a managed node, the node registers this problem in one of several possible ways.

Integrating examples

You can integrate examples into primary modules (for example, definition lists, procedures, processes, topics, and troubleshooting scenarios) and secondary modules (for example, itemized lists and notes).

Examples in definition lists

You can add examples to definition list items. To avoid overwhelming definition list items, do not add long examples.

Change	To

`Protocol`
Type of transfer protocol used.
`Type`
Type of source document.
	`Protocol`
	Type of transfer protocol used *(for example, HTTP).*
	`Type`
	Type of source document *(for example, SGML or HTML).*

`$OPC_MSG_IDS` Returns the message IDs of the messages currently selected in the message browser.	`$OPC_MSG_IDS` Returns the message IDs of the messages currently selected in the message browser. *Sample output:* *`85432efa-ab4a-71d0-14d4-0f887a7c0000`*

TIP For related guidelines, see "Definition lists" on page 63.

Examples in itemized lists

You can add examples to itemized list items. To avoid overwhelming itemized list items, do not add long examples.

Change	To
▪ **Popup menus** If you right-click an item, a popup menu appears. From there, you can choose menu items with keyboard shortcuts.	▪ **Popup menus** If you right-click an item, a popup menu appears. From there, you can choose menu items with keyboard shortcuts. *For example, if you want to create a filtered active message browser on a particular node, right-click the node, then press A.*

TIP For related guidelines, see "Itemized lists" on page 98.

Examples in notes

You can add examples to notes, tips, cautions, and warnings. To avoid overwhelming these small text blocks, do not add long examples.

Change	To
CAUTION: If you set the refresh interval too high, you may not be informed of status changes in a timely manner.	**CAUTION:** If you set the refresh interval too high, you may not be informed of status changes in a timely manner. *For example, if you set the refresh interval to 24:00:00, you receive notifications only once a day.*

TIP For related guidelines, see "Notes" on page 106.

Examples in procedures

You can add examples to procedure steps. Make sure to emphasize the procedure steps, not the examples.

Change	To
1 Click `[Browse]` to select a web browser *(for example, ICEbrowser)* installed on your computer.	1 Click `[Browse]` to select a web browser installed on your computer. *For example, you could select ICEbrowser.*

TIP For related guidelines, see "Procedures" on page 118.

Examples in processes

You can add examples to process steps. Make sure to emphasize the process steps, not the examples.

Change	To
1 **Detecting problems** AcmePro enables you to detect problems in the message browser, review node status, and direct problem management activities.	1 **Detecting problems** AcmePro enables you to detect problems in the message browser, review node status, and direct problem management activities. *For example, if you notice that a node icon has changed to red (critical), scan the message browsers for messages with the status color red.*

TIP For related guidelines, see "Processes" on page 124.

Examples in topics

You can add simple or complex examples to topics. If a complex example takes on a life of its own, convert the example into a new topic.

Change	To
The Web Browser Properties dialog box looks different, depending on the type of web browser you have configured.	The Web Browser Properties dialog box looks different, depending on the type of web browser you have configured *(for example, Netscape Communicator)*.
Node icons are displayed in colors that correspond to the highest severity of any message received from that node.	Node icons are displayed in colors that correspond to the highest severity of any message received from that node. *For example, the node icon turns red when the affected node has at least one unacknowledged critical message.*

For example, a stylesheet named `docbook.xsl` might contain a template rule for `literal` elements.	For example, a stylesheet named `docbook.xsl` might contain a template rule for `literal` elements:
	<xsl:template match="literal">
	    <pre>
	    <xsl:apply-templates/>
	    </pre>
	</xsl:template>

TIP For related guidelines, see the following sections:

- "Labeling examples" on page 88
- "Topics" on page 140

Examples in troubleshooting scenarios

You can add examples to problem descriptions and solution steps. In solutions, make sure to emphasize the steps, not the examples.

Change	To
Problem The color scheme on your system has identical background and foreground colors. Or there is not enough color contrast between the foreground and background. This problem is more common on UNIX than on other operating systems.	**Problem** The color scheme on your system has identical background and foreground colors. Or there is not enough color contrast between the foreground and background. This problem is more common on UNIX than on other operating systems. *For example, on Solaris, you might experience problems with the NorthernSky color scheme.*
Solution Do the following: 1 Make sure the management server is running. From an MS-DOS terminal window, enter the ping command: **ping <*management_server*>** 2 ...	**Solution** Do the following: 1 Make sure the management server is running. From an MS-DOS terminal window, enter the ping command: **ping <*management_server*>** *For example, enter the following:* ***ping wintermute.ai.com*** 2 ...

TIP For related guidelines, see "Troubleshooting scenarios" on page 144.

Figures

Figures are images or charts. When introducing figures, use standard wording and electronic cross-references. If you are building print and online documents from a single source, optimize images for print, then convert them to online formats. When embedding images in documents, reference the images rather than copying them into the documents.

TIP For related guidelines, see the following sections:

- "Captions" on page 51
- "Cross-referencing figures" on page 56

Types of figures

Figures can be images, charts, or both:

- **Images**

  Realistic graphics that can be referenced in documents (for example, a screenshot of a software program).

- **Charts**

  Abstract graphics that can be drawn with lines, arrows, circles, squares, and so on (for example, bar charts, pie charts, flow charts, and so on).

One common way to combine images and charts is to annotate screenshots with text and arrows.

Introducing figures

When introducing figures, use standard wording and electronic cross-references. Reference figure numbers in print, and figure titles online.

Print	Online
Once you have entered your new password, you receive a system message confirming your password change, as shown in *Figure 1-3*. Figure 1-3 Password change confirmation	Once you have entered your new password, you receive a system message confirming your password change, as shown in Figure: Password change confirmation. Figure: Password change confirmation

Optimizing images

Whenever possible, optimize figures for print and online viewing:

- **Print**

  For print viewing, save images in TIFF (Tagged Image File Format) format at an acceptable print resolution, which is normally 300 or 600 dots per inch (dpi). If possible, save the images in full color, so they can be converted to any online format.

- **Online**

  For online viewing, save images in GIF (Graphics Interchange Format) or JPEG (Joint Photographic Experts Group) format at web browser resolution, which is 72 dpi. GIF format is best for images that contain text (for example, screenshots). JPEG format is best for images that do not contain text (for example, photographs).

When building print and online documents from a single source, optimize images for print, then convert them to online formats.

Parameter	Print	Online	Screenshots	Photographs
Color	Full	Indexed	✓	
		Full		✓
Format	TIFF	GIF	✓	
		JPEG		✓
Resolution (dpi)	300 or 600	72	✓	✓

TIP Image processing applications (for example, Adobe Photoshop and Equilibrium DeBabelizer) enable you to optimize TIFF, GIF, and JPEG images.

Embedding figures

When building a single-source document, you can copy images into the document, or you can reference images from the document:

- **Copied images**

  If you copy an image into a single-source document, you cannot edit the image. To re-use the image in another part of the same document, you need to recopy the image into the document again, thereby increasing the size of the document.

- **Referenced images**

  If you reference an image from a single-source document, you can reference the image from another part of the same document without increasing the size of the document. If you edit the image with a graphics application (for example, Adobe Photoshop), your changes are automatically reflected wherever the image is referenced.

When embedding images in a single-source document, reference the images rather than copying them into the document. You can then use document conversion programs to convert the images to other formats automatically.

TIP Some authoring tools (for example, Adobe FrameMaker) enable you to reference the same image in many documents. If you reference TIFF images in your print documents, you can use a conversion program (for example, Quadralay WebWorks Publisher) to convert the referenced TIFF images to GIF or JPEG format when you convert the print files to an online format.

For more about development tools, see "Development tools" on page 188.

Integrating figures

You can integrate images or charts into procedures, processes, topics, and troubleshooting scenarios. However, images and charts can overwhelm the procedures, processes, and solution steps they illustrate. For this reason, consider adding figures to topics only, then cross-referencing the figures from procedures, process, and solution steps. If you do add a figure to a procedure, a process, or a troubleshooting scenario, consider adding the figure at the end of the module, then cross-referencing the figure from the relevant step.

Change	To
To print application output	**To print application output**
To print application output, follow these steps:	To print application output, follow these steps:
1 To make sure the application output window is active, click it.	1 To make sure the application output window is active, click it.
2 From the File menu, select `Print`. The Print dialog box opens, *as shown in Figure 1-2.* [Figure 1-2] This dialog box varies, depending on which version of JRE is installed on your computer.	2 From the File menu, select `Print`. The Print dialog box opens, *as shown in Figure 1-2 on page 25.*
3 Click `[Print]`. The application output is printed directly to the default printer.	3 Click `[Print]`. The application output is printed directly to the default printer.
	About the Print dialog box The Print dialog box *is shown in Figure 1-2.* [Figure 1-2] This dialog box varies, depending on which version of JRE is installed on your computer.

Change	To
To view assigned applications To view all applications assigned to you, do one of the following: ▪ **Menu bar** In the Actions menu, select `Start`. *Figure 1-1 shows the menu bar.* [Figure 1-1] ▪ **Object pane** In the object pane, double-click an application. *Figure 1-2 shows the object pane.* [Figure 1-2]	**About the menu bar** The menu bar is the strip of pulldown menus at the very top of the Java GUI, *as shown in Figure 1-1.* [Figure 1-1] **About the object pane** The object pane is the second pane below the toolbar and position controls, *as shown in Figure 1-2.* [Figure 1-2] **To view assigned applications** To view all applications assigned to you, do one of the following: ▪ **Menu bar** In the Actions menu, select `Start`. *Figure 1-1 on page 24 shows the menu bar.* ▪ **Object pane** In the object pane, double-click an application. *Figure 1-2 on page 25 shows the object pane.*

TIP For related guidelines, see the following sections:

- "Procedures" on page 118
- "Processes" on page 124
- "Topics" on page 140
- "Troubleshooting scenarios" on page 144

Front matter

Front matter is material that appears at the front of a book, before the content begins. Although front matter has its origin in print media, it has parallels in online media. You can re-use (repurpose and re-assemble) front matter from print documents in online documents, and vice versa.

TIP For more information about repurposing and re-assembly, see "Types of single sourcing" on page 15.

Types of front matter

Front matter consists of document-based modules that appear before content-based modules:

- **Print**

  In print documents, document-based title pages, copyright pages, tables of contents, prefaces, and introductions appear before content-based chapters and appendices.

- **Online**

  In online documents, document-based welcome pages, legal notice pages, and tables of contents usually appear before content-based topics, procedures, definitions, and so on.

No matter what the format, front matter is associated with a particular output document, not single-source input modules.

NOTE In print documents, appendices, glossaries, and indexes are referred to as "back matter." For single sourcing, this category is less than helpful because it groups content-based modules (appendices) with document-based modules (glossaries and indexes). In single sourcing, tables of contents, section contents, inline cross-references, and indexes are document-based cross-references. For more information about cross-referencing documents, see Step 7, "Cross-referencing documents," on page 38.

Repurposing front matter

To repurpose front matter in print and online documents, look for ways to use the same modules in both media. For example, most print and online documents have a copyright page. Rather than building two pages, use a single-source copyright page for both.

Print	Online
Title page	—
Copyright page	Copyright page
Table of contents	Table of contents
Preface	Welcome page
Introduction	—

Re-assembling front matter

To re-assemble front matter, you can use conditional text to mark some sections for one type of document, and other sections for another type of document. For example, in the preface to a printed manual, you can mark print sections with `PrintOnly` conditional text, and online sections with `OnlineOnly` conditional text. When you convert the single-source preface to an online help format, you can switch on the `OnlineOnly` conditional text, and switch off the `PrintOnly` conditional text. When you print the manual, you can do the opposite.

Print	Online
Preface This guide introduces you to AcmePro SE for Linux....	**Welcome** Welcome to AcmePro SE for Linux Online Help. To view any topic, click the topic title in the left window frame.

TIP For more about conditional text, see "Conditional text" on page 183.

Glossaries

Glossaries are alphabetized definition lists that explain technical terms found in a document. These terms can be proprietary or public domain terminology related to the product you are documenting. In glossary terms, use abbreviations rather than spellouts. Spell out the abbreviations at the beginning of their definitions. Begin definitions with nouns or verbs, not articles. Do not use a term to define itself. Develop master glossaries to ensure consistent terms and definitions, and to eliminate redundant work.

TIP For related guidelines, see the following sections:

- "Capitalizing glossaries" on page 156
- "Cross-referencing glossaries" on page 57
- "Definition lists" on page 63
- "Labeling glossaries" on page 89
- "Parallel glossary entries" on page 159

Types of glossary terms

Glossaries are definition lists that define technical terms used in documents:

- **Proprietary terms (primary)**

  The primary goal of a glossary is to define proprietary terms (for example, "Windows Registry") related to the product you are documenting.

- **Public domain terms (secondary)**

  The secondary goal of a glossary is to define non-proprietary terms (for example, "TCP/IP") related to the product you are documenting.

Include proprietary and public domain terms in your glossary.

TIP To help users find definitions wherever they are in your documents, you can link technical terms in modules to their glossary definitions. In printed documents, these terms are often italicized. In online documents, the terms are hyperlinked to glossary definitions. To help writers use consistent terms and definitions, you can build master glossaries that include all terms and definitions used in all documents related to a given product. For more about master glossaries, see "Building master glossaries" on page 83.

Building glossary terms

In glossary terms, use abbreviations, rather than their spellouts. Spell out the abbreviations at the beginning of their definitions.

Change	To
document type definition Definition of a document type in SGML or XML.	**document type definition** *See* DTD. **DTD** document type definition. Definition of a document type in SGML or XML.

Building glossary definitions

In glossary definitions, answer a simple question: "What is it?" Begin definitions with nouns or verbs, not articles. Do not use a term to define itself. Define the term with a synonym, and use the synonym throughout the definition. To eliminate redundant terms, organize multiple definitions of the same term as numbered definitions.

Change	To
DocBook The DocBook DTD is a system for writing structured documents with SGML or XML.	**DocBook** DTD for writing structured documents with SGML or XML.
application (function) Simple script, process, or command. **application (product)** Complex product that includes a number of programs and configuration files.	**application** 1. Simple script, process, or command. 2. Complex product that includes a number of programs and configuration files.

Building master glossaries

A master glossary is a shared glossary for related documents or document sets. Master glossaries enable users to find definitions to technical terms found in many different documents. They also enable you to eliminate redundancy and inconsistency in your terms and definitions.

Types of master glossaries

There are two types of master glossaries:

- **External use**

  You can compile all the glossaries found in all the documents in a given document set. You can then attach this master glossary to all documents within the document set.

- **Internal use**

  You can compile all the glossaries found in all the documents your project team, department, division, or company produces. You can then re-use this internal "database" for future projects.

Compiling master glossaries

Compiling a master glossary is a cyclical process:

1. **Organize**

   Organize individual glossaries into one single-source file. Often, compilation is a simple (if tedious) copy-and-paste operation.

2. **Edit**

   Edit the single-source glossary thoroughly. Make sure the terms and definitions are consistent.

3. **Update**

   Ask everyone on your team to provide corrections as well as new terms and definitions. Incorporate these changes into the master glossary.

Repeat this process, as needed.

Re-using external master glossaries

If you develop a master glossary for a particular document set, you can add the complete master glossary to each document in the set. For example, if you are using Adobe FrameMaker, you can add the master glossary file to each book file in your document set.

How you re-use master glossary terms and definitions is often determined by your local tools, technologies, and traditions. No matter how you re-use your master glossary, make sure to maintain the single-source glossary in a central location that all team members can access.

Re-using internal master glossaries

If you develop a master glossary for internal use only, you can re-use terms and definitions in many different ways, such as the following:

- **Importing**
  You can import selected terms and definitions from the master glossary into individual glossaries. For example, if you are using Adobe FrameMaker+SGML, you can import selected DocBook `GlossEntry` elements.
- **Exporting**
  You can use conditional text to mark terms and definitions in the master glossary for a particular document. You can then switch on only that conditional text style, and save the master glossary as a glossary for your target document. For example, you could mark selected terms and definitions with `UserGuide` conditional text, save the file as a new glossary file, and add the new glossary file to your user guide book file.

TIP For related guidelines, see the following sections:

- "About authoring tools" on page 189
- "Conditional text" on page 183

Headings

Headings are titles to primary modules, which are assembled as document sections. When building headings, answer specific questions as directly as possible. Use a consistent heading syntax that indicates the module type, and exactly what information the module contains. Make sure the headings are understandable when taken out of context.

TIP For related guidelines, see the following sections:

- "Capitalizing headings" on page 157
- "Captions" on page 51
- "Parallel headings" on page 160

Headings as answers

Use headings to answer specific questions as directly as possible.

Question	Heading
Who should use AcmePro?	Who should use AcmePro
What does AcmePro do?	What AcmePro does
When should I upgrade AcmePro?	When to upgrade AcmePro
Where is the AcmePro executable?	Location of important files
Why should I use AcmePro?	Reasons for using AcmePro
How do I install AcmePro?	To install AcmePro

TIP One easy and effective way to answer the question directly is to use a heading syntax that mirrors the syntax of the question itself.

Labeling modules

When labeling modules, use a consistent and predictable heading syntax that indicates the module type, and exactly what information the module contains.

Module type	Heading syntax
Definitions	Types of dialog boxes
Definition 1	About the ABC dialog box
Definition 2	About the DEF dialog box
Procedures	Installing AcmePro
Procedure 1	To install AcmePro automatically
Procedure 2	To install AcmePro manually
Processes	Problem solving
Process 1	Detecting problems
Process 2	Investigating problems
Process 3	Solving problems
Process 4	Documenting solutions
Topics	About Acme Pro
Topic 1	Who should use AcmePro
Topic 2	What AcmePro does
Topic 3	Reasons for using AcmePro

TIP When setting up a heading syntax for different module types, you can use the first word of the heading to indicate the module type. For example, use "About..." for general descriptions, "Types of..." for lists, "To..." for procedures, "If..." for troubleshooting scenarios, and so on. This approach enables users to scan for particular types of information in your documents.

Which heading syntax you follow is less important than consistency. The more consistent your heading syntax, the more predictable your headings become. The more predictable your headings, the easier it is for users to find exactly the information they need.

Labeling definition lists

When labeling definition lists, use a consistent heading syntax that distinguishes different types of definitions.

Definition list type	Heading syntax
Applications	Types of applications
Application 1	About Cancel Reboot
Application 2	About Job Status
Application 3	About Memory Load
Commands	Types of commands
Command 1	About the backup command
Command 2	About the restore command
Libraries	Types of libraries
File 1	About the include file
File 2	About the make file
Menus	Types of menus
Menu 1	About the File menu
Menu 2	About the Edit menu
Variables	Types of variables
Variable set 1	About console variables
Variable set 2	About SNMP variables

TIP For related guidelines, see "Definition lists" on page 63.

Labeling examples

Most examples do not have headings. Simple examples, which appear within parentheses, obviously cannot have headings. Complex examples, which are formatted as paragraphs, sometimes do have headings. Within topics, introduce long examples with a word or phrase, and a colon. If the example is itself a topic, add a descriptive heading.

Example within a topic

About anchor pseudo-classes

The anchor (`A`) element uses pseudo-classes, which show the status of hyperlinks in HTML pages. You can set style rules for each pseudo-class.

Example:

```
A:link { color: blue }

A:visited { color: red }

A:active { color: black }
```

Example as a topic

Sample HTML page

If converted to online help, "About anchor pseudo-classes" on page 23 would require the following HTML:

```
<html>
<head>
<title>About anchor pseudo-
classes</title>
</head>
<body>
<h1>About anchor pseudo-
classes</h1>
<p>The anchor (<code>A</
code>) element uses pseudo-
classes, which show the
status of hyperlinks in HTML
pages. You can set style
rules for each pseudo-
class.</p>
<p>Example:</p>
<pre>
A:link { color: blue }
A:visited { color: red }
A:active { color: black }
</pre>
</body>
</html>
```

TIP For related guidelines, see "Examples" on page 67.

Labeling glossaries

When labeling glossary divisions, use "Symbols" for symbolic characters, and use "Numerals" for numeric characters. In alphabetic labels, never use more than one character. Ignore missing alphabetic characters.

Change	To
A	**Numerals**
32-bit application Application that works with information 32 bits at a time.	32-bit application Application that works with information 32 bits at a time.
H, I, J	J
JavaScript Cross-platform scripting language developed for use on the Internet.	JavaScript Cross-platform scripting language developed for use on the Internet.

TIP For related guidelines, see "Glossaries" on page 81.

Labeling itemized lists

When labeling itemized lists, use a consistent and predictable heading syntax that distinguishes different types of information (for example, document conventions, installation requirements, and so on).

Itemized list type	Heading syntax
Requirements Requirement 1 Requirement 2	Installation requirements Hardware requirements Software requirements
Section contents	In this chapter In this appendix

TIP For related guidelines, see "Itemized lists" on page 98.

Labeling procedures

When labeling procedures, answer the question "How?" with a verb. Use a heading syntax that distinguishes superprocedures from other procedures. For superprocedures, use gerunds (that is, verbs ending in "ing") and plural nouns. For subprocedures, multiple-step procedures, and single-step procedures, use action verbs and singular nouns (that is, "To...").

Procedure type	Heading syntax
Superprocedure	Installing AcmePro
Subprocedure 1	To install AcmePro on Linux
Subprocedure 1	To install AcmePro on Windows
Multiple-step procedure	To configure AcmePro on UNIX
Single-step procedure	To de-install AcmePro

TIP For related guidelines, see "Procedures" on page 118.

Labeling processes

When labeling processes, answer the question "How?" with a verb. As a rule, use gerunds (that is, verbs ending in "ing") rather than action verbs.

Process type	Heading syntax
Processes	Building documents
Process 1	Identifying modules
Process 2	Labeling modules
Process 3	Organizing modules

TIP For related guidelines, see "Processes" on page 124.

Labeling topics

When labeling topics, answer specific questions. Use a consistent syntax that distinguishes different types of topics.

Topic type	Heading syntax
Concepts	About AcmePro
Description 1	What is AcmePro?
Description 2	Who should use AcmePro
Description 3	What AcmePro does
Argument 1	Reasons for using AcmePro

TIP For related guidelines, see "Topics" on page 140.

Labeling troubleshooting scenarios

When labeling troubleshooting scenarios, describe problems, not their solutions. Write the headings from the user perspective, rather than from the system perspective. To describe problems, use explicit hypothetical statements (for example, "If you cannot start Acme Pro") or implicit hypothetical statements (for example, "Cannot start AcmePro"). Whenever possible, use parallel construction.

Change	To
Troubleshooting	Troubleshooting runtime problems
Failure when starting AcmePro	Cannot start AcmePro
Application startup problem	Cannot start an application
Commands don't execute	Cannot execute commands

TIP For related guidelines, see "Troubleshooting scenarios" on page 144.

Indexes

Indexes are alphabetized cross-reference lists that help users locate information quickly. Wherever possible, nest index entries. Format index entries for print and online formats. To help users find information in document sets, build master indexes.

TIP For guidelines to indexing print and online documentation, refer to *Indexing: A Nuts-and-Bolts Guide for Technical Writers* by Kurt Ament. This guide is available from William Andrew Publishing.

Types of index formats

You can format indexes for print or online delivery:

- **Print indexes**

  In print indexes, you link index entries to content modules with page numbers. In page numbers, include start and end ranges, so users can see at a glance the size of the referenced section.

- **Online indexes**

  In online indexes, you link index entries to content modules with hyperlinks. In hyperlinks, you include start ranges only.

Whether you format indexes for print documents, online documents, or both, it is wise to build index entries with no more than three levels. Index entries with more than three levels are difficult to use in print documents, and almost impossible to use in online documents.

TIP Building a master index is good preparation for single sourcing because it forces you to establish standards for individual indexes, which you then combine into shared indexes. The better your indexing guidelines, the easier it is to re-use the individual index entries in other documents. For details, see "Building master indexes" on page 95.

Types of indexing tools

There are two types of indexing tools:

- **Integrated tools**

  With an indexing tool that is integrated into an authoring tool (for example, Adobe FrameMaker), you add hyperlinked index markers to individual modules. You then generate indexes automatically.

- **Stand-alone tools**

  With a stand-alone indexing tool (for example, CINDEX), you build indexes manually to match a particular format. These manual indexes are extremely difficult to maintain, let alone re-use in other formats.

Integrated indexing tools are ideal for single sourcing because they enable you to generate indexes dynamically, then convert the generated indexes into other formats automatically.

TIP For more about authoring tools, see "About authoring tools" on page 189.

Building index hierarchies

If two or more index entries begin with the same first word or phrase, nest the entries under that common word or phrase.

Change	To

position controls, 147
showing position controls, 148
showing window panes, 151–153
window panes, 150
	position controls
	    description, 147
	    showing, 148
	showing
	    position controls, 148
	    window panes, 151–153
	window panes
	    description, 150
	    showing, 151–153

TIP For related guidelines, see "Parallel index entries" on page 161.

Formatting index entries

Format index entries for output formats. For print, include start and end ranges. For online, include start ranges only. Use document conversion programs to convert print formats to online formats, or vice versa.

Print	Online
acknowledgements	acknowledgements
See also acknowledging messages	*See also* acknowledging messages
automatic, 123	automatic
description, 135-136	description
reviewing, 136	reviewing
acknowledging messages	acknowledging messages
See also acknowledgements	*See also* acknowledgements
all messages, 155	all messages
selected messages, 156	selected messages

Shortening index entries

Shorten index entries for online formats by removing non-essential words (for example, articles and prepositions). In particular, online help systems provide very little space in which to view indexes.

Change	To
printing a document, 3	printing document, 3

Windows
    compatibility with, 23
    metering applications in, 24
	Windows
	    compatibility, 23
	    metering applications, 24

TIP Online help systems that provide small windows for document content force you to break information into very small chunks that fit their windows. By forcing you to break information into small chunks, online help systems actually teach you how to improve the usability of print documents.

Building master indexes

A master index is an index shared by related documents. Master indexes enable users to find answers across document sets. Master indexes also enable you to eliminate redundancy and inconsistency in document sets.

When to use master indexes

Normally, you build a master index only for documents that are delivered together as a document set. These shared indexes are extremely valuable to users who are looking for an answer to a specific question, but do not know which document in the reference set contains the answer. For example, if you are documenting an enterprise solution comprised of five different products explained in five different documents, it may be wise to build a master index. A master index enables users to find answers without first choosing between one of five different document indexes every time they have a question. If your documents are online, the master index helps you build seamless cognitive bridges between them.

Master indexes often help writers as much as users. Combining individual indexes into a master index can help you diagnose structural problems with your document set. For example, if your index has five identical index entries, this redundancy indicates a problem. Either the master index is redundant, or the documents are redundant.

Master indexes can also help you avoid duplicate work. For example, if you want to use the same modules in two different documents with two incompatible index structures, you may be tempted to use conditional text to develop two sets of index tags, one for each document. Rather than doing duplicate work, and most likely introducing errors, you would be well advised to build a master index shared by the two documents.

TIP For more about conditional text, see "Conditional text" on page 183.

Compiling master indexes

If you are using an authoring tool that enables you to embed index markers in individual files (for example, Adobe FrameMaker), you can build a master book file that contains all the single-source files in your document set. You can then generate the master index from the master book file.

CAUTION If you attach a master index to books in a reference set, you need to change the page numbers of the index file itself for each book. To do so, generate the master index, then regenerate the table of contents (but *not* the index) for each separate book.

When compiling master indexes, make sure to do the following:

- **Abbreviate document titles in page numbers**
  To make it clear which documents are related to which index entries, include abbreviated document titles as prefixes to page numbers (for example, LUG-25). If possible, add the prefixes in a master book file, rather than in individual source files.
- **Spell out abbreviations in page footers**
  To explain document title abbreviations in page numbers, include a key in each footer of the master index (for example, LUG = *AcmePro for Linux User's Guide*). If possible, use variables for document titles, and add the footers to master pages.
- **Add primary product names**
  To make it clear which products are associated with which entries, add product names (for example, "AcmePro SE for Linux") to individual index entries. If possible, add product names to index markers so you can regenerate the master index whenever the content changes.

TIP Although you add abbreviations to the master index document, you set up the master index document to read page numbers from individual files. For online formats, you do not have to worry about page numbers or page footers because index entries are simply hyperlinks. However, if you think you may someday use the master index for print formats, add abbreviated document titles to page numbers. Later, you can strip these page-number prefixes from index entries with your conversion templates.

Editing master indexes

Once you compile individual indexes into a master index, you eliminate redundancies and inconsistencies by renesting entries. If a given entry has more than one set of page numbers associated with it, you break down the entry, pinpointing the difference between the two sections in the two documents. The more consistently you nest the individual indexes, the less you have to renest entries in the master index.

Change	To
printing, CG-25 to CG-28, UG-27 to UG-30	printing description, CG-25 to CG-28 procedure, UG-27 to UG-30
installation requirements, A-15 to A-18, B-17 to B-20	installation requirements Product A, A-15 to A-18 Product B, B-17 to B-20

TIP Often, there are conflicts between individual indexes and master indexes. By editing index tags *for* the master index, you edit the index tags *against* individual indexes. Although it is possible to develop "compromise" entries, the "path to peace" can be extremely complex. To protect your sanity, develop individual indexes or master indexes, but not both.

If possible, edit index tags embedded in individual documents, not the master index itself. If you edit index tags, you can regenerate the master index whenever content changes.

Single-source indexing reduces your localization costs as well. For details, see "Localization" on page 195.

Itemized lists

Itemized lists display serial items vertically for easy scanning. Normally, each list item is associated with a small, text-based graphical element (for example, a square, a disc, or an em dash).

TIP For related guidelines, see the following sections:

- "Active voice in itemized lists" on page 178
- "Capitalizing itemized lists" on page 157
- "Labeling itemized lists" on page 89
- "Parallel list items" on page 162
- "Section contents" on page 126

Types of itemized lists

You can build simple or complex itemized lists:

- **Simple lists**

  Itemized lists that do not contain sublists or annotation. These lists display serial items that could be contained within a sentence. They enables users to scan short items, rather than read long sentences.

- **Complex lists**

  Itemized lists that include sublists, annotations, or both:

  - *Sublists*

    Second-level lists nested under first-level lists. Or third-level lists nested under second-level lists.

  - *Annotations*

    Paragraphs that explain list items. You can use paragraph breaks to separate list items and annotations.

  Graphical elements (for example, squares, discs, and em dashes) help distinguish first-, second-, and third-level list items.

You can use itemized lists as stand-alone modules. Or you can integrate itemized lists into other modules.

TIP If you find serial words or phrases separated by commas in a sentence, consider formatting the words or phrases as itemized lists. For details, see "Formatting sentences as lists" on page 172.

Introducing itemized lists

When introducing itemized lists, target the lists. Separate introductions from itemized lists. Punctuate the introductions with a colon (:).

Change	To
AcmePro enables you to create database reports using the report writer that is supplied with the installed database or report-writing tool. *You can do the following with database reports.* ▪ Display reports in a window ▪ Save reports to file ▪ Print reports	AcmePro enables you to create database reports using the report writer that is supplied with the installed database or report-writing tool. *With database reports, you can do the following:* ▪ Display reports in a window ▪ Save reports to file ▪ Print reports

Wording list items

Use parallel construction for all items in an itemized list. For example, begin all items with a verb or a noun. Likewise, if one list item forms a complete sentence, add a period to all items in the list.

Change	To
In optional mode, only the owner of a message may do the following: ▪ Perform operator-initiated *actions* related to the message. ▪ You can *escalate* the message. ▪ Message *acknowledgement*	In optional mode, only the owner of a message may perform the following: ▪ **Actions** Perform operator-initiated actions related to the message. ▪ **Escalations** Escalate the message. ▪ **Acknowledgements** Acknowledge the message.

TIP For related guidelines, see "Parallel list items" on page 162.

Formatting list items

When formatting list items, establish list styles. Emphasize primary items.

Establishing list styles

Establish list styles for first-, second-, and third-level list items.

For example, you could use the following list styles:

- **First level**
  Squares (▪) for first-level list items.
- **Second level**
  Discs (•) for second-level list items.
- **Third level**
  Em dashes (–) for third-level list items.

Use the list styles consistently.

<table>
<tr><th>Change</th><th>To</th></tr>
<tr><td>To view all applications assigned to you, do one of the following:
▪ In the Actions menu, select one of the following:
<code>Start</code>
<code>Start Customized</code>
▪ In the object pane, do one of the following:
Double-click an application in the Applications folder.
Right-click an application in the Applications folder, then select one of the following from the popup menu:
<code>Start</code>
<code>Start Customized</code></td><td>To view all applications assigned to you, do one of the following:
▪ Menu bar
In the Actions menu, select one of the following:
• <code>Start</code>
• <code>Start Customized</code>
▪ Object pane
In the Applications folder, do one of the following:
• Double click
Double-click an application.
• Right click
Right-click an application. Then, from the popup menu, select one of the following:
– <code>Start</code>
– <code>Start Customized</code></td></tr>
</table>

Emphasizing primary items

In complex lists, use bold or italic type to emphasize primary list items. Special type distinguishes these list "headings" from the annotations and sublists nested beneath them.

<table>
<tr><th>Change</th><th>To</th></tr>
<tr><td>The main window of the Java GUI is divided into four main areas:

▪ Shortcut bar in the top-left pane provides you with shortcuts to frequently used objects. For details, see "About the shortcut bar" on page 5.

▪ Object pane in the top-middle pane displays the structure of your managed environment. For details, see "About the object pane" on page 7.

▪ Workspace pane in the top-right pane contains multiple tabs, each containing one workspace. For details, see "About the workspace pane" on page 15.

▪ Browser pane in the bottom pane provides you with quick access to the latest messages. For details, see "About the browser pane" on page 27.</td><td>The main window of the Java GUI is divided into four main areas:

▪ Shortcut bar
Top-left pane provides you with shortcuts to frequently used objects.
For details, see "About the shortcut bar" on page 5.

▪ Object pane
Top-middle pane displays the structure of your managed environment.
For details, see "About the object pane" on page 7.

▪ Workspace pane
Top-right pane contains multiple tabs, each containing one workspace.
For details, see "About the workspace pane" on page 15.

▪ Browser pane
Bottom pane provides you with quick access to the latest messages.
For details, see "About the browser pane" on page 27.</td></tr>
</table>

Integrating itemized lists

You can integrate itemized lists into definition lists, procedures, processes, section contents, topics, and troubleshooting scenarios.

Itemized lists in definition lists

You can use itemized lists for serial items within definitions.

Change	To
`Print` The Print menu item prints selected messages, all messages in the message browser, details of selected messages, or details of all messages in the message browser.	`Print` Enables you to print the following types of messages from the message browser: • All messages • Selected messages • Details of all messages • Details of selected messages

TIP For related guidelines, see "Definition lists" on page 63.

Itemized lists in procedures

You can use itemized lists for options within procedure steps. You can also use itemized lists for single-step procedures.

Change	To
To view selected messages To view selected messages, follow these steps: 1 In the object pane, right-click a node, message group, or service. A popup menu appears. 2 ...	**To view selected messages** To view selected messages, follow these steps: 1 In the object pane, right-click one of the following: • Message group • Node • Service A popup menu appears. 2 ...

Change	To
To view all active messages To view all active messages, select `Filtering > All Active Messages` in the Actions menu. A message browser opens in the workspace pane, displaying all active messages.	**To view all active messages** To view all active messages, do this: ▪ In the Actions menu, select `Filtering > All Active Messages.` A message browser opens in the workspace pane, displaying all active messages.

TIP For related guidelines, see "Procedures" on page 118.

Itemized lists in processes

You can use itemized lists for options within process steps.

Change	To
Solving problems The problem solving process includes the following steps: 1 **Identify the problem.** An error message indicates where and why the problem occurred. 2 **Solve the problem.** To solve the problem, you can use automatic actions, broadcast commands, or operator-initiated actions.	**Solving problems** The problem solving process includes the following steps: 1 **Identify the problem.** An error message indicates where and why the problem occurred. 2 **Solve the problem.** To solve the problem, you can use any of the following tools: • Automatic actions • Broadcast commands • Operator-initiated actions

TIP For related guidelines, see "Processes" on page 124.

Itemized lists in section contents

You can use annotated itemized lists to build section contents. For long sections, add annotations and cross-references to the lists.

Change	To
This section describes how to view all active messages, history messages, pending messages, and selected messages.	This section describes how to view the following: ▪ All active messages ▪ History messages ▪ Pending messages ▪ Selected messages
This appendix explains how to enter menu bar and popup menu commands from the keyboard without using the mouse; options contained in the menu bar; icons in the toolbar; popup menu items; and options that appear within dialog boxes.	This appendix contains the following sections: ▪ **About keyboard shortcuts** Explains how to enter menu bar and popup menu commands from the keyboard without using the mouse. ▪ **Types of menu bar items** Explains options contained in the menu bar. ▪ **About toolbar icons** Explains icons in the toolbar. ▪ **Types of popup menu items** Explains popup menu items. ▪ **Types of dialog boxes** Explains options that appear within dialog boxes.

TIP For related guidelines, see "Section contents" on page 126.

Itemized lists in topics

You can add itemized lists to topics. If a sentence contains many serial items, consider converting the sentence into an itemized list.

Change	To
AcmePro contains three navigation tools. The menu bar is the top row of pulldown menus. The toolbar is the row of icons just below the menu bar. The position controls are the narrow band of horizontal arrows just below the toolbar.	AcmePro contains the following navigation tools: ▪ **Menu bar** Top row of pulldown menus. ▪ **Toolbar** Row of icons just below the menu bar. ▪ **Position controls** Narrow band of horizontal arrows just below the toolbar.

TIP For related guidelines, see "Topics" on page 140.

Itemized lists in troubleshooting scenarios

You can use itemized lists for options within troubleshooting solutions.

Change	To
Solution To prevent the problem from recurring, change the related template condition. Or fix the script or command.	**Solution** To prevent the problem from recurring, do one of the following: ▪ Change the related template condition. ▪ Fix the script or command.

TIP For related guidelines, see "Troubleshooting scenarios" on page 144.

Notes

Notes are small text blocks that supplement other modules. These text blocks offer positive advice (notes and tips) or negative advice (cautions and warnings) to users. Such advice appears within a given module, and comments on that module. The advice is often highlighted with a box or an icon to attract attention.

NOTE The word "note" has a general meaning and a specific meaning. Generally, "note" refers to all types of advisories (notes, tips, cautions, and warnings). Specifically, "note" denotes a specific type of advisory (notes only). This guide uses "note" in both senses. This double usage is industry standard.

Types of notes

Most product documentation contains four different types of notes:

- **Note**
  Neutral or positive clarification that applies only in special cases, or that qualifies important points.
- **Tip**
  Positive suggestion that helps users apply information described in a module to meet their specific needs.
- **Caution**
  Negative alert that a particular action can result in data loss, data corruption, security problems, or performance problems.
- **Warning**
  Negative alert that a particular action can result in physical harm to users, or can result in damage to hardware.

TIP These types of notes are industry standard. Avoid vague note constructions (for example, "IMPORTANT"). Use a note, tip, caution, or warning instead.

Building notes and tips

Use notes for clarifications that apply in special cases, or that qualify important points. Use tips for positive suggestions that help users apply information described in a module to meet their specific needs.

Note

To add a shortcut group to the shortcut bar, follow these steps:

1. Right-click the shortcut group.

   NOTE: The new group is added immediately below the currently selected shortcut group.

2. From the popup menu, select `Add New Group`.
3. In the Add New Group dialog box, add a shortcut group label.
4. Click `[OK]`.

Tip

To search for a specific item in the object pane, follow these steps:

1. In the Edit menu, select `Find`.

   The Find dialog box appears.

2. In the Find dialog box, enter your search criteria.

   TIP: For descriptions of options in the Find dialog box, see "About the Find dialog box" on page 23.

3. Click `[Find Next]`.

Building cautions and warnings

Use cautions for negative alerts that a particular action can result in data loss, data corruption, security problems, or performance problems. Use warnings for negative alerts that a particular action can result in physical harm to users, or can result in damage to hardware.

Caution

CAUTION: If your administrator has changed your responsibilities, managed nodes, or applications, you must reload your configuration manually. Otherwise, you cannot see the new configuration. For details, see <u>To reload the configuration</u>.

Warning

WARNING: Failure to follow these instructions may result in serious or fatal injury.

Integrating notes

You can integrate notes, tips, cautions, and warnings into definition lists, procedures, processes, topics, and troubleshooting scenarios. Make sure these secondary modules are left-aligned with the paragraphs, list items, or steps to which they refer. Never "stack" notes, tips, cautions, or warnings.

Notes in definition lists

In definition lists, make sure notes, tips, cautions, and warnings are left-aligned with the list items to which they refer.

Change	To
`Create Filter On Message` Opens the Filter Messages dialog box. **TIP:** This option is a good starting point for creating a message filter.	`Create Filter On Message` Opens the Filter Messages dialog box. **TIP:** This option is a good starting point for creating a message filter.

TIP For related guidelines, see "Definition lists" on page 63.

Notes in procedures

In procedures, make sure notes, tips, cautions, and warnings are left-aligned with the steps to which they refer.

Change	To
To add a new shortcut to the shortcut bar, follow these steps: 1 Right-click a shortcut. **NOTE:** The new shortcut is added immediately below the currently selected shortcut. 2 ...	To add a new shortcut to the shortcut bar, follow these steps: 1 Right-click a shortcut. **NOTE:** The new shortcut is added immediately below the currently selected shortcut. 2 ...

TIP For related guidelines, see "Procedures" on page 118.

Notes in processes

In processes, make sure notes, tips, cautions, and warnings are left-aligned with the steps to which they refer.

Change	To
The problem solving process includes the following steps: 1 **Detecting problems** The most common indication of a problem in your managed environment is an error message in your message browser. **TIP:** For more about error messages, see "About error messages" on page 94. 2 ...	The problem solving process includes the following steps: 1 **Detecting problems** The most common indication of a problem in your managed environment is an error message in your message browser. **TIP:** For more about error messages, see "About error messages" on page 94. 2 ...

TIP For related guidelines, see "Processes" on page 124.

Notes in topics

In topics, make sure notes, tips, cautions, and warnings are left-aligned with the text to which they refer.

Change	To
Colors show you which messages are related to a which severity level. **TIP:** For a description of message colors, see "About message colors" on page 69.	Colors show you which messages are related to a which severity level. **TIP:** For a description of message colors, see "About message colors" on page 69.

TIP For related guidelines, see "Topics" on page 140.

Notes in troubleshooting scenarios

In troubleshooting scenarios, make sure notes, tips, cautions, and warnings are left-aligned with the steps to which they refer.

Change	To
Solution Do the following: 1 Make sure the management server is running. From an MS-DOS terminal window, enter the `ping` command: **ping <*management_server*>** 2 Make sure the server processes are running on the management server. NOTE: Ask your administrator to check the status of the server processes.	**Solution** Do the following: 1 Make sure the management server is running. From an MS-DOS terminal window, enter the `ping` command: **ping <*management_server*>** 2 Make sure the server processes are running on the management server. NOTE: Ask your administrator to check the status of the server processes.

TIP For related guidelines, see "Troubleshooting scenarios" on page 144.

Serial notes

Never "stack" notes, tips, cautions, or warnings. Either combine them, or separate them with text.

Change	To
TIP: For a description of default elements in the message headline, see "About the message browser headline" on page 337. **TIP:** For a description of each message attribute and flag, see "Types of message attributes" on page 339.	**TIP:** For a description of default elements in the message headline, see "About the message browser headline" on page 337. For a description of each message attribute and flag, see "Types of message attributes" on page 339.

Organization

Document organization drives document assembly. When organizing modular content in single-source input documents, integrate secondary modules into primary modules. Then segregate primary modules by type. When organizing (assembling) output documents, use logical hierarchies to target specific audiences and purposes. Then flatten the hierarchies to pull information to the surface of the output documents.

TIP One of the best ways to test your document organization is to generate a detailed table of contents that shows all the sections and subsections in your document. If you do not want your final table of contents to show all the subsections in your document, generate interim tables of contents for testing purposes. For details, see "Tables of contents" on page 138.

Organizing input modules

When organizing single-source input modules, you integrate secondary modules, then segregate primary modules:

- **Integrating secondary modules**

  Secondary modules are "helper" modules (for example, figures, itemized lists, notes, and tables) that supplement primary modules. Integrate secondary modules into primary modules.

  For details, see "Integrating secondary modules" on page 30.

- **Segregating primary modules**

  Primary modules (for example, definition lists, glossary definitions, procedures, processes, topics, and troubleshooting scenarios) embody distinct types of information. Segregate primary modules into distinct sections of your single-source document.

  For details, see "Segregating primary modules" on page 31.

TIP Even when organizing output documents, it is wise to separate primary modules by type. For example, never integrate a definition list into a procedure. Instead, separate the two modules into distinct sections, then cross-reference them. For details, see "Cross-references" on page 54.

Organizing output documents

You can organize document sections and subsections by alphabetical order, type of audience, level of detail, level of importance, physical location, sequential order, or type of module.

You can combine these organizational strategies as well. For example, you could group procedures (type), order them in a logical sequence (time), group subprocedures with their parent superprocedures (detail), and introduce installation and de-installation procedures first (importance).

Organizing by alphabet

You can organize modules alphabetically by heading. For example, you can order procedures and definition lists alphabetically, for easy reference.

Sequence	Alphabet
Chapter 2. Getting started	**Chapter 2. Getting started**
To edit a document	To edit a document
To save a document	To print a document
To print a document	To save a document

TIP Alphabetical organization is surprisingly usable, especially if headings follow a predictable syntax. For example, if you order procedures alphabetically, users can find the procedure they need quickly.
For more about heading syntax, see "Headings" on page 85.

Organizing by audience

You can organize modules by user. For example, you can separate operator and administrator sections.

Operator	Administrator
Chapter 1. About agents	**Chapter 2. Installing agents**
What are agents?	To install the agent
What do agents do?	To activate the managed node
How do agents work?	To set an IP alias for the agent

Organizing by detail

You can organize modules by level of detail. For example, you can introduce general product features before explaining specific product functions.

Specific to general	General to specific
Chapter 1. About AcmePro	**Chapter 1. About AcmePro**
How to order AcmePro	What is AcmePro?
How AcmePro works	How AcmePro works
What is AcmePro?	How to order AcmePro

Organizing by importance

You can organize modules by level of importance. For example, you can list installation and de-installation procedures before other procedures.

Sequence	Importance
Chapter 2. Getting started	**Chapter 2. Getting started**
To install AcmePro	To install AcmePro
To configure AcmePro	To de-install AcmePro
To de-install AcmePro	To configure AcmePro

Organizing by location

You can organize modules by the physical location of the objects they describe. For example, you can introduce the parts of an application in the order in which they appear in the user interface.

Alphabet	Location
Appendix A. About pulldown menus	**Appendix A. About pulldown menus**
About the Edit menu	About the File menu
About the File menu	About the Edit menu
About the View Menu	About the View Menu

Organizing by sequence

Organize modules in sequential order. For example, you can include installation procedures before configuration procedures.

Importance	Sequence
Chapter 2. Getting started	**Chapter 2. Getting started**
To install AcmePro	To install AcmePro
To de-install AcmePro	To configure AcmePro
To configure AcmePro	To de-install AcmePro

Organizing by type

You can organize modules by type, segregating primary modules into distinct sections. Document sections that are segregated by module type are extremely usable because they make it easy for users to find exactly the type of information they need (for example, a particular procedure).

Location	Type
Chapter 1. Using the shortcut bar	**Chapter 1. About AcmePro**
About the shortcut bar	About the shortcut bar
About shortcut bar menus	About the object pane
To customize the shortcut bar	About the workspace pane
Chapter 2. Using the object pane	**Chapter 2. Customizing AcmePro**
About the object pane	To customize the shortcut bar
About object pane menus	To customize the object pane
To customize the object pane	To customize the workspace pane
Chapter 3. Using the workspace pane	**Chapter 3. Types of AcmePro menus**
About the workspace pane	About shortcut bar menus
About workspace pane menus	About object pane menus
To customize the workspace pane	About workspace pane menus

TIP If you organize documents by module type, make sure to cross-reference related modules. For details, see "Cross-references" on page 54.

Flattening hierarchies

Complex document hierarchies help writers, but hinder users:

- **Writers**
  Documents with many different heading levels help writers understand products. Writers can logically group many different levels of information into many different heading levels. Complex hierarchies help writers build coherent documents.
- **Users**
  Documents with many different heading levels can actually hinder users from using products. Many different heading levels force users to think about your document rather than use the products it explains. The moment users are forced to think, they stop using. As a writer, your goal is to decrease user downtime, not increase it.

In this conflict between you and your users, always choose your users. To make complex documents appear simple to your users, reduce the number of heading levels. By flattening your document hierarchies in this way, you pull information to the surface of the document, where it "pops out" at users. In other words, you make the document more usable.

To flatten documents, you can promote sections to chapters or appendices. Or you can promote subsections to sections.

TIP Complex hierarchies are especially unusable in online documents, which do not provide users with enough visual space to see complex hierarchies. Typically, online help systems provide users with an electronic window that is about half the size of the printed page. As a result, more than three heading levels are extremely difficult for users to recognize online. Because online formats constitute your worst-case scenario, test hierarchies on online documents before testing them on print documents. For more about usability testing, see Step 9, "Testing documents," on page 42.

Promoting sections

An easy way to simplify complex hierarchies is to promote sections. For example, if you want to simplify a chapter with many different heading levels, you can convert each first-level section to a chapter. In so doing, you convert one complex hierarchy into many simple hierarchies.

Change	To
Chapter 1. Problem solving	PART I. PROBLEM SOLVING
Detecting problems	**Chapter 1. Detecting problems**
Monitoring messages	Monitoring messages
Viewing messages	Viewing messages
...	...
Investigating problems	**Chapter 2. Investigating problems**
Investigating message histories	Investigating message histories
Investigating pending messages	Investigating pending messages
...	...
Solving problems	**Chapter 3. Solving problems**
Verifying automatic actions	Verifying automatic actions
Verifying operator actions	Verifying operator actions
...	...
Documenting solutions	**Chapter 4. Documenting solutions**
Acknowledging messages	Acknowledging messages
Annotating messages	Annotating messages
...	...

NOTE If you promote many sections to chapters, add document parts (divisions) to separate different types of chapters.

Promoting subsections

A difficult way to simplify hierarchies is to promote subsections. For example, if you want to simplify a chapter with many different heading levels, you can convert each second-level heading to a first-level heading.

Change	To
Chapter 1. About AcmePro	**Chapter 1. About AcmePro**
Tour of AcmePro	Tour of AcmePro
Shortcut bar	About the shortcut bar
Object pane	About the object pane
Workspace pane	About the workspace pane
Browser pane	About the browser pane
Types of message browsers	Types of message browsers
Active message browser	About the active message browser
History message browser	About the history message browser
Pending messages browser	About the pending messages browser
Types of popup menus	Types of popup menus
Shortcut bar popup menu	About the shortcut bar popup menu
Object pane popup menu	About the object pane popup menu
Workspace pane popup menu	About the workspace pane popup menu
Browser pane popup menu	About the browser pane popup menu

CAUTION Promoting subsections can sometimes make it difficult for users to see the relationships between sections and what were previously their subsections. Promote subsections only if you cannot promote sections (for example, convert first-level sections to chapters).

Before you promote subsections, make sure to build section contents from sections to subsections. For details, see "Section contents" on page 126.

Procedures

Procedures are step-by-step instructions that explain how to perform tasks. There are four types of procedures: single-step procedures, multiple-step procedures, superprocedures, and subprocedures. Regardless of the type, introduce all procedures with standard wording. Use procedure steps to emphasize user actions and options.

TIP For related guidelines, see the following sections:

- "Active voice in procedures" on page 178
- "Cross-referencing procedures" on page 60
- "Labeling procedures" on page 90
- "Parallel procedure steps" on page 163
- "Present tense in procedures" on page 174

Types of procedures

There are four types of procedures:

- **Single-step procedures**

  Stand-alone procedures that contain only one distinct user action, or step. Format this step as an itemized list with one list item.
- **Multiple-step procedures**

  Stand-alone procedures that contain more than one distinct user action, or step. Format these steps as ordered (numbered) list items.
- **Superprocedures**

  Procedures that contain other procedures. If a procedure contains more than nine steps, convert it into a superprocedure. Cross-reference each step in the superprocedure to one subprocedure.
- **Subprocedures**

  Procedures that explain the steps in a superprocedure. Cross-reference each subprocedure from the appropriate step in its superprocedure.

In each type of procedure, steps represent distinct user actions.

Introducing procedures

When introducing procedures, use standard wording. Begin introductions with the goal of the procedure. End introductions by pointing to the procedure steps. Punctuate procedure introductions with colons (:).

Change	To
Do the following tasks in the specified order to install AcmePro successfully.	To install AcmePro, follow these steps:

Emphasizing single steps

To emphasize single-step procedures, use itemized lists. An itemized list item makes it clear where the introduction ends and the procedure begins.

Change	To
Following is the procedure for forcing shared memory modules to be loaded automatically. Enter **`forceload: sys/shmsys`** at the end of the `/etc/system` file.	To force shared memory modules to load automatically, do this: ▪ Enter the following lines at the end of the `/etc/system` file: **`forceload: sys/shmsys`**

Emphasizing substeps

To emphasize substeps in procedures, use sublists to make each substep stand out.

Change	To
1 Right-click the message, then select `Node in Object Pane` from the popup menu. This highlights the node in the object pane.	1 Highlight a node in the object pane: a Right-click a message. A popup menu appears. b In the popup menu, select `Node in Object Pane`.

Emphasizing long steps

In long procedures with large amounts of text between steps, emphasize each step with bold or italicized type to make the step stand out.

Change	To
To integrate a new threshold monitor, follow these steps: 1 Place the monitor, script, or executable in the monitor directory. For Solaris managed nodes, you can place monitor programs or scripts in the following directory on the management server: `/var/opt/AP/share/databases/OpC/mgd_node/customer/\ s700/solaris/monitor` 2 Configure the threshold monitor template. Use the following windows: • *Add Threshold Monitor* Use this window to configure a threshold monitor template. • *Message Templates* Use this window to review, add, modify, or delete threshold monitor templates. Each template defines a monitor.	To integrate a new threshold monitor, follow these steps: 1 **Place the monitor, script, or executable in the monitor directory.** For Solaris managed nodes, you can place monitor programs or monitor scripts in the following directory on the management server: `/var/opt/AP/share/databases/OpC/mgd_node/customer/\ s700/solaris/monitor` 2 **Configure the threshold monitor template.** Use the following windows: • *Add Threshold Monitor* Use this window to configure a threshold monitor template. • *Message Templates* Use this window to review, add, modify, or delete threshold monitor templates. Each template defines a monitor.

Emphasizing keyboard input

Remove redundant information from instructions involving the keyboard. For example, "holding down" a key assumes you have already pressed it.

Change	To
Press and hold down the **Ctrl** key and the **Alt** key, then press the **Del** key.	Press **Ctrl+Alt+Del**.

Emphasizing mouse input

Remove redundant information from instructions involving the mouse. On personal computers, the default mouse button is the left button. If you use the word "click," users assume "left-click." However, if you want users to click something with the right mouse button, use the word "right-click."

Change	To
Use the left mouse button to select the conditions you require.	*Click* the conditions you need.
In the object pane, *click the right mouse button over* a managed node, message group, or service.	In the object pane, *right-click* one of the following: ▪ Managed node ▪ Message group ▪ Service

Emphasizing command-line input

When instructing users to enter commands from the command line, aim the sentence in the direction of the command. Explicitly instruct users to "enter" commands. Separate command descriptions into definition lists.

Change

To start the DCE daemon from the command line

Do the following to start the DCE daemon from a terminal window without using the System Management Interface Tool (SMIT) GUI:

```
mkdce -a <user> -n <cell> -s <sec_server> rpc
```

The following variables are contained in this command:

<user>
: Name of the privileged user (for example, `cell_adm`).

<cell>
: Name of the cell (for example, `apo`).

<sec_server>
: Name of the system that is configured as the master security server.

To

To start the DCE daemon from the command line

You can start the DCE daemon from a terminal window without using the System Management Interface Tool (SMIT) GUI.

To start the DCE daemon from the command line, do this:

- Enter the following command:

  ```
  mkdce -a <user> -n <cell> -s <sec_server> rpc
  ```

  For descriptions of `mkdce` variables, see "About the mkdce command" on page 437.

About the mkdce command

The `mkdce` command contains the following variables:

<user>
: Name of the privileged user (for example, `cell_adm`).

<cell>
: Name of the cell (for example, `apo`).

<sec_server>
: Name of the system that is configured as the master security server.

TIP For related guidelines, see "Definition lists" on page 63.

Emphasizing options within steps

Within procedure steps, use itemized lists for optional actions.

Change	To
1 In the Filter Messages dialog box, select all or specified pending messages. Click [OK] to see all pending messages. To see only some pending messages, use the General tab or the Symbols and Objects tab.	1 In the Filter Messages dialog box, select one of the following: • *All pending messages* To see all pending messages, click [OK]. • *Specified pending messages* To see only some pending messages, use the following tabs: — General — Symbols and Objects

TIP For related guidelines, see "Itemized lists" on page 98.

Emphasizing optional steps

Begin optional steps with a clear visual and verbal indication (for example "*Optional:*") that they are not mandatory.

Change	To
4 Optionally, you can print the contents of the Message Properties window. To start the print job, click [Print]. The Print dialog box opens. There you can specify further options.	4 *Optional:* Print the contents of the Message Properties window. To start the print job, click [Print]. The Print dialog box opens. There you can specify further options.

Processes

Processes are similar to procedures and topics. Like procedures, processes answer one question: How? Because these answers are sequential, they are often formatted as ordered lists. Like topics, processes are narratives rather than imperatives. They explain what someone or something does (declarative), not what users should do (imperative).

TIP For related guidelines, see the following sections:

- "Active voice in processes" on page 179
- "Labeling processes" on page 90
- "Parallel process steps" on page 164
- "Present tense in processes" on page 175

Writing narrative processes

When writing narrative processes, use active voice and present tense.

Change	To
Problem solving There are three steps in the problem solving process. First, you are notified exactly when and where a problem occurs, including where and why the problem occurred, and what you should do to resolve the problem. Second, the following tools are available to help you solve problems: automatic actions, broadcast commands, and operator-initiated actions. Finally, to document solutions, you use message annotations to record how a problem was resolved, and then acknowledge the message to remove it from the current work area, storing it in the history database.	**Problem solving** The problem solving process includes four steps. First, when a problem occurs, an error message notifies you. This message explains exactly where, when, and why the problem occurred. And it suggests steps you can take to solve the problem. Second, you solve the problem with one of the following system tools: an automatic action, a broadcast command, or an operator-initiated action. Third, you add a message annotation to record how the problem was resolved. Finally, you acknowledge the message to move it from the current work area to the history database.

Building process steps

Whenever possible, organize processes into sequential steps. As in procedures, emphasize steps and options. Separate process steps from details about those steps. Include itemized lists or substeps, as needed.

Change

Problem solving

The problem solving process includes four steps. First, when a problem occurs, an error message notifies you. This message explains exactly where, when, and why the problem occurred. And it suggests steps you can take to solve the problem. Second, you solve the problem with one of the following system tools: an automatic action, a broadcast command, or an operator-initiated action. Third, you add a message annotation to record how the problem was resolved. Finally, you acknowledge the message to move it from the current work area to the history database.

To

Problem solving

The problem solving process includes the following steps:

1 **Detecting problems**

   When a problem occurs, an error message notifies you. The message explains exactly where, when, and why the problem occurred. And it suggests steps you can take to solve the problem.

2 **Solving problems**

   To solve the problem, you can use any of the following tools:

   - Automatic actions
   - Broadcast commands
   - Operator-initiated actions

3 **Documenting solutions**

   To record how you solved the problem, you add a message annotation.

4 **Closing problems**

   You close the original error message to move it from the current work area to the history database.

TIP For related guidelines, see the following sections:

- "Itemized lists" on page 98
- "Procedures" on page 118

Section contents

Section contents are lists that introduce subsections. These lists are to sections what tables of contents are to documents. Tables of contents summarize document structure. Section contents summarize section structure. Section contents are particularly important in online documents, where they supplement the main navigation system by hyperlinking document sections to their subsections.

Although section contents can appear in many different formats, the most common format is an itemized list. For long sections, add annotations that describe what each subsection contains. To re-use section contents for print and online documents, include electronic cross-references.

TIP For related guidelines, see the following sections:

- "Cross-references" on page 54
- "Indexes" on page 92
- "Itemized lists" on page 98
- "Tables of contents" on page 138

Introducing small sections

When introducing small sections that contain subsections, use itemized lists rather than sentences.

Change	To
This section describes the File, Edit, and View menus.	This section describes the menu items that appear in the menu bar: ■ File menu ■ Edit menu ■ View menu

Introducing large sections

When introducing large sections (for example, chapters or appendices), use complex lists. Include list "headings" (highlighted text), as well as annotations that indicate what each subsection contains.

Change	To
This appendix describes options that appear in the menu bar, popup menus, and dialog boxes.	This appendix describes options that appear in the user interface: ▪ **Menu bar options** Describes options that appear in the menu bar. ▪ **Popup menu options** Describes options that appear in popup menus. ▪ **Dialog box options** Describes options that appear in dialog boxes.

Cross-referencing section contents

When cross-referencing section contents in print documents, add page numbers. When cross-referencing section contents in online documents, add hyperlinks.

Print	Online
This section describes the menu items that appear in the menu bar: ▪ "About the File menu" on page 4 ▪ "About the Edit menu" on page 6 ▪ "About the View menu" on page 8	This section describes the menu items that appear in the menu bar: ▪ About the File menu ▪ About the Edit menu ▪ About the View menu

Print	Online
This appendix describes options that appear in the user interface:	This appendix describes options that appear in the user interface:
▪ **Menu bar options** Describes options that appear in the menu bar. For details, see "Types of menu bar items" on page 11.	▪ **Menu bar options** Describes options that appear in the menu bar. For details, see Types of menu bar items.
▪ **Popup menu options** Describes options that appear in popup menus. For details, see "Types of popup menu items" on page 22.	▪ **Popup menu options** Describes options that appear in popup menus. For details, see Types of popup menu items.
▪ **Dialog box options** Describes options that appear in dialog boxes. For details, see "Types of dialog boxes" on page 33.	▪ **Dialog box options** Describes options that appear in dialog boxes. For details, see Types of dialog boxes.

TIP Using electronic hyperlinks enables you to convert cross-references to different print and online formats. Electronic cross-references also automate updates to document contents, section contents, inline cross-references, and indexes. For details, see "Cross-references" on page 54.

With high-end authoring tools (for example, Adobe FrameMaker), you can use conditional text to switch on section contents for online documents, then switch off section contents for printed documents. For more about conditional text, see "Conditional text" on page 183.

Tables

Tables are collections of columns and rows that compare information in a small visual space. By presenting information more concisely than text or figures, tables enable at-a-glance comparison. Use tables only when comparing items. Do not use tables where you could use lists. That is, do not format definition lists, itemized lists, procedures, or processes as tables. When introducing tables, use standard wording and electronic cross-references. To eliminate redundant table cells, combine redundant rows and columns.

TIP For related guidelines, see the following sections:

- "Captions" on page 51
- "Cross-referencing tables" on page 60
- "Parallel table cells" on page 166

When to use tables

As a rule, you should use tables only when comparing items.

Tables are two-dimensional lists that often behave like images:

- **Verbal lists**

  Tables list information. These lists display text, images, or both, in two dimensions. They are similar to definition lists and itemized lists.

- **Visual maps**

  Tables map information. These maps enable users to compare different types of information at a glance. They behave like charts.

By combining the strengths of verbal and visual communication, tables enable users to compare complex clusters of information quickly.

TIP For related guidelines, see the following sections:

- "Definition lists" on page 63
- "Itemized lists" on page 98

When not to use tables

If you use tables when you could use definition lists or itemized lists, you can actually cause usability and re-usability problems:

- **Usability problems**

  Given their visual complexity, tables can sometimes make information look more complicated than it actually is. For example, formatting a glossary as a two-column table with borders adds visual "noise" to the glossary, thereby making it more difficult to scan.

- **Re-usability problems**

  Because table content is tied to a complex format of columns and rows, converting tables from one document format to another is often more complicated than converting nested definition lists and itemized lists.

Normally, tables should have three or more columns. If tables have only two columns, you should think about converting them to definition lists or itemized lists.

NOTE This guide uses two-column tables to compare examples, side by side. Although it would be possible to format the examples as definition lists or itemized lists, the side-by-side table format makes comparison easier.

Introducing tables

When introducing tables, use standard wording and electronic cross-references. In print documents, cross-reference table numbers, not titles. If the table appears on a different page than the cross-reference, include the page number in the cross-reference. In online documents, table captions are not normally numbered. Cross-reference the table titles.

Change	To
The table on page 198 shows the SMS services and their associated monitors. [Table A-3]	*Table A-3 on page 198* shows the SMS services and their associated monitors. [Table A-3]

NOTE The example tables in this guide do not have captions. Because the example tables in this guide are so closely related to the guidelines they illustrate, the section headings serve as captions for the tables. More often than not, adding table numbers and titles would be redundant. Instead of cross-referencing table captions, this guide cross-references section titles, which summarize guidelines and examples alike.

Avoiding tabular lists

Do not use tables where you could use lists. That is, do not format definition lists, itemized lists, procedures, or processes as tables.

Avoiding tabular definition lists

Do not use tables where you could use definition lists. When formatted as tables, lists appear more complicated than they actually are. Definition lists are ideal for explaining product components and terminology. When formatted as definition list items, terms and definitions are easy to scan. Make sure to include a heading and introduction for each definition list.

Change

Table 1-1 shows the options contained in the View menu.

Table 1-1 View menu options

Item	Definition
`Shortcut Bar`	Shows or hides the shortcut bar.
`Object Pane`	Shows or hides the object pane.

To

About the View menu

The View menu contains the following options:

`Shortcut Bar`
: Shows or hides the shortcut bar.

`Object Pane`
: Shows or hides the object pane.

TIP For related guidelines, see "Definition lists" on page 63.

Avoiding tabular itemized lists

Do not use tables where you could use itemized lists. When formatted as tables, itemized lists appear more complicated than they actually are. Itemized lists are ideal for presenting serial information. When formatted as itemized list items, serial information is easy to scan.

Change

Table 1-2 shows the problems you can investigate with the message browser.

Table 1-2 Investigating problems with the message browser

Problem	Description
Basic information	You get basic information about a message by reviewing the message browser headline in the workspace pane or the browser pane.
Detailed information	You get detailed information about a message by reviewing the Message Properties dialog box.

To

Investigating problems with the message browser

You can use the message browser to investigate the following problems:

- **Basic information**

  You get basic information about a message by reviewing the message browser headline in the workspace pane or the browser pane.

- **Detailed Information**

  You get detailed information about a message by reviewing the Message Properties dialog box.

TIP For related guidelines, see "Itemized lists" on page 98.

Avoiding tabular procedures

Do not use tables for procedures. When formatted as tables, procedures appear more complicated than they actually are. Ordered lists are ideal for presenting sequential procedure steps. When formatted as ordered list items, procedure steps are easy to scan.

Change

Table 1-3 shows how to acknowledge a message.

Table 1-3 Acknowledging a message

Step	Description
1	In the message browser, select the message you want to acknowledge.
2	From the Actions menu, select `Acknowledge Messages`.

To

To acknowledge a message

To acknowledge a message, follow these steps:

1. In the message browser, select the message you want to acknowledge.
2. From the Actions menu, select `Acknowledge Messages`.

TIP For related guidelines, see "Procedures" on page 118.

Avoiding tabular processes

Do not use tables for processes. When formatted as tables, processes appear more complicated than they actually are. Ordered lists are ideal for presenting sequential process steps. When formatted as ordered list items, process steps are easy to scan.

Change

Table 1-4 shows the problem solving process.

Table 1-4 Problem solving process

Step	Description
1	Detect the problem.
2	Investigate the problem.
3	Solve the problem.
4	Document the solution.

To

Solving problems

The AcmePro problem solving process includes the following steps:

1. Detect the problem.
2. Investigate the problem.
3. Solve the problem.
4. Document the solution.

TIP For related guidelines, see "Processes" on page 124.

Eliminating redundant table cells

Tables enable users to compare complex clusters of information at a glance. Like charts, tables communicate visually. Charts and tables map information in a way that enables users to see the forest as well as the trees. In effect, tables and charts simplify complexity. They present complex information in a way that appears deceptively simple.

One of the most effective ways to make complex information appear simple is to eliminate redundant information. To eliminate redundant information in tables, combine redundant rows and columns.

TIP For related guidelines, see "Parallel table cells" on page 166.

Combining redundant rows

To eliminate redundant rows, combine the rows. By combining redundant rows, you build visual hierarchies within tables. These visual hierarchies enable users to identify "parent" and "child" information at a glance.

Change

Table 1-5 Redundant rows

Categories	Items
Category 1	Item 1.1
Category 1	Item 1.2
Category 2	Item 2.1
Category 2	Item 2.2

To

Table 1-5 Combined rows

Categories	Items
Category 1	Item 1.1
	Item 1.2
Category 2	Item 2.1
	Item 2.2

Combining redundant columns

To eliminate redundant columns, combine the columns. By combining redundant columns, you build "subtables" within tables. These subtables enable users to compare distinct groups of complex information simultaneously.

Change

Table 1-6 Redundant columns

Categories	Items
Group 1	Group 1
Category 1.1	Item 1.1.1
	Item 1.1.2
Category 1.2	Item 1.2.1
	Item 1.2.2
Group 2	Group 2
Category 2.1	Item 2.1.1
	Item 2.1.2
Category 2.2	Item 2.2.1
	Item 2.2.2

To

Table 1-6 Combined columns

Categories	Items
Group 1	
Category 1.1	Item 1.1.1
	Item 1.1.2
Category 1.2	Item 1.2.1
	Item 1.2.2
Group 2	
Category 2.1	Item 2.1.1
	Item 2.1.2
Category 2.2	Item 2.2.1
	Item 2.2.2

Integrating tables into topics

You can integrate tables into topics. As a rule, do not add tables to other types of primary modules (for example, procedures).

<table>
<tr><th>Change</th><th>To</th></tr>
<tr><td rowspan="2">To add an application group to the shortcut bar

To add an application group to the shortcut bar, follow these steps:

1 Right-click the application group immediately above the place where you want to add a new shortcut group.

2 From the popup menu, select <code>Add New Application Group</code>.
The Add New Application Group dialog box appears.

3 From the input text field, add a shortcut group label.

4 Click <code>[OK]</code>.
The application group is added to the shortcut bar, as shown in Table 1-1.
[Table 1-1]</td><td>Creating new application groups

You can create new application groups that are added individually at the end of the shortcut bar. The same tree level found under the application group is added to the shortcut bar, as shown in Table 1-1.
[Table 1-1]</td></tr>
<tr><td>To add an application group to the shortcut bar

To add an application group to the shortcut bar, follow these steps:

1 Right-click the application group immediately above the place where you want to add a new shortcut group.

2 From the popup menu, select <code>Add New Application Group</code>.
The Add New Application Group dialog box appears.

3 From the input text field, add a shortcut group label.

4 Click <code>[OK]</code>.
The application group is added to the shortcut bar, as shown in Table 1-1 on page 27.</td></tr>
</table>

TIP For related guidelines, see "Topics" on page 140.

Tables of contents

Tables of contents link the modules that make up a document. No matter what their format, all print or online documents require some sort of table of contents. Although a table of contents can take many forms, it always shows the structure of the assembled document.

TIP For related guidelines, see the following sections:

- "Cross-references" on page 54
- "Headings" on page 85
- "Indexes" on page 92
- "Section contents" on page 126

Levels of detail

In your table of contents, you can provide users with summary or detailed views of your document:

- **Summary**

  To provide users with a quick overview of your document, you can include only one or two levels of headings in the table of contents. By withholding heading levels, you make the document appear simple.

- **Details**

  To provide users with a detailed view of your document, you can include all sections and subsections in the table of contents. By including all sections and subsections, you make it easy for users to find particular subsections.

Each approach has its strengths and weaknesses. Choose a level of detail that matches the audience, purpose, and format of your document.

Even if you plan to include a table of content with a summary view in your final document, you can generate a table of contents with a detailed view in interim drafts to verify your heading structure.

TIP In HTML-based online documentation, you can develop dynamic tables of contents users can expand or contract by mouse click or rollover. To develop these flexible tables of contents, you can use Dynamic HTML (DHTML), a combination of JavaScript and Cascading Style Sheet (CSS) coding.

Targeting tables of contents

In your table of contents, provide users with a summary or detailed view of your document, depending on your audience, purpose, and format.

Summary view	Detailed view
About this guide	**About this guide** Audience Purpose ...

Chapter 1. About Acme Pro
In this chapter
About dashboards
...
	Chapter 1. About Acme Pro
	In this chapter
	About dashboards
	    About the Message dashboard
	    About the Diagnostic dashboard
	    ...
	...

Appendix A. About AcmePro options
In this appendix
Types of menus
...
	Appendix A. About AcmePro options
	In this appendix
	Types of menus
	    About the File menu
	    About the Edit menu
	    ...
	...

Glossary **Index**	**Glossary** **Index**

TIP If you provide a summary view, make sure to include section contents in each major section of your document. For more about section contents, see "Section contents" on page 126.

Topics

Topics are texts that answer specific questions. To answer these questions, you can use arguments, descriptions, expositions, or narrations. In most product documentation, the majority of topics are descriptions. To make your thoughts clear, order sentences and paragraphs by level of importance and detail. To make your sentences clear, speak directly to users in active voice, second person, and present tense.

TIP For related guidelines, see the following sections:

- "Active voice in processes" on page 179
- "Cross-referencing topics" on page 61
- "Labeling topics" on page 91
- "Present tense in topics" on page 176

Types of topic questions

Topics are texts that answer one of the following questions:

- **Who?**
  Example: Who should use AcmePro?
- **What?**
  Example: What is AcmePro?
- **When?**
  Example: When should I upgrade AcmePro?
- **Where?**
  Example: Where do I find the AcmePro executable?
- **Why?**
  Example: Why upgrade AcmePro?

Each topic should answer one basic question.

TIP In contrast to topics, procedures always explain *how* to perform tasks. For details, see "Procedures" on page 118.

Types of topic answers

To answer specific questions, you can use one of the following writing methods:

- **Argument**
  Convincing users to do something. For example, you could advise users to restart their computers after installation. Arguments are often included in notes, tips, cautions, and warnings.
- **Description**
  Presenting objects or situations in a way that enables users to visualize them. For example, you could describe a graphical user interface. Descriptions are often accompanied by figures.
- **Exposition**
  Presenting facts or ideas logically. For example, you could present the installation requirements for a program. Expository writing is often accompanied by itemized lists, definition lists, or tables.
- **Narration**
  Presenting a series of events in sequential or chronological order. For example, you could explain an installation process. Narrations can also be presented as ordered lists.

In technical writing, the vast majority of topics are descriptions. That is, most of the time, topics describe what things are, or what they do.

TIP To make clear to users what type of answers to expect in topics, you should follow a consistent heading syntax. Consistent headings add predictability to your topics. For details, see "Headings" on page 85.

Organizing topics clearly

In topics, order sentences and paragraphs to maximize your message.

- **Importance**
  Order sentences and paragraphs by level of importance. Answer important questions first, then answer less important questions. For example, explain what a product is before you explain what it does.
- **Detail**
  Order sentences and paragraphs by level of detail. Answer general questions first, then answer specific questions. For example, introduce general product features before explaining specific product functions.

Change	To
About AcmePro Secure	**About AcmePro Secure**
Advanced Network Security (ANS) is a software bundle that provides protection, based on the secret-key protocol for data exchanged between the AcmePro management server and the managed nodes. Generic Security Service (GSS) is a customizable security package that provides a wide variety of encryption algorithms and protocols. Secure SSL (SSSL) is an optional feature that provides a secure link to the management server through an SSL implementation. All three products are included in an add-on package to AcmePro. This security add-on package is called AcmePro Secure.	AcmePro Secure is a security add-on package to AcmePro. AcmePro Secure contains Advanced Network Security (ANS), Generic Security Service (GSS), and Secure SSL (SSSL). ANS is a software bundle that provides protection, based on the secret-key protocol for data exchanged between the AcmePro management server and the managed nodes. GSS is a customizable security package that provides a wide variety of encryption algorithms and protocols. SSSL is an optional feature that provides a secure link to the management server through an SSL implementation.

TIP When structuring topics, your goal is to minimize surprise. The best way to minimize surprise is to structure topics that answer big questions first. To answer big questions first, you can start each paragraph of a given topic with a sentence that answers a general question. Then finish each paragraph with more and more details that support the answer.

The first sentence of each paragraph can serve as a "headline." These headlines provide users with quick summaries of the paragraphs. The summaries make it much easier for users to scan topics for precisely the information they need.

If you consistently provide users with topics that are easy to scan, your document becomes predictable. That is, users can predict where to find different levels of answers. Predictable documents earn user trust.

Writing topics clearly

Whenever writing topics, speak directly to your users. Wherever possible, use active voice, second person, and present tense.

Change	To
Customers using a version of AcmePro prior to 5.0 must first upgrade to 5.0 before *they* can upgrade to 6.0.	If *you* are using a version of AcmePro lower than 5.0, *you* must first upgrade to 5.0 before upgrading to 6.0.
An AcmePro installation *should be carefully planned*.	*Plan* your AcmePro installation carefully.
Close the FTP connection when the files *have been transferred* successfully.	Once the files *transfer* successfully, close the FTP connection.

TIP For related guidelines, see the following sections:

- "Person" on page 167
- "Tense" on page 173
- "Voice" on page 177

Troubleshooting scenarios

Troubleshooting scenarios are a hybrid of topics and procedures. Introduce each scenario by summarizing one problem and its solution. Then explain the problem and the solution separately and in detail. When describing a problem, focus on the problem, not its cause. Format solutions as procedures. Group closely related problems and solutions.

TIP For related guidelines, see the following sections:

- "Cross-referencing troubleshooting scenarios" on page 62
- "Labeling troubleshooting scenarios" on page 91
- "Procedures" on page 118
- "Topics" on page 140

About problems and solutions

Troubleshooting scenarios are a hybrid of topics and procedures:

- **Problems (topics)**

  Troubleshooting scenarios begin with problems descriptions. Typically, these descriptions are written as descriptive topics.

- **Solutions (procedures)**

  Troubleshooting scenarios end with problem solutions. Typically, these solutions are written as step-by-step procedures.

As a rule, each troubleshooting scenario should focus on one problem and one solution only.

TIP Occasionally, problems are so closely related that it makes sense to include them in one troubleshooting scenario. On these occasions, separate each distinct problem and solution within the module. For details, see "Grouping related problems and solutions" on page 147.

Introducing troubleshooting scenarios

Begin troubleshooting scenarios by introducing problems and solutions together. Begin introductions with "If..." statements.

Change	To
Cannot start AcmePro After entering the login information, you get an error box with the following message: `Management server is not running.`	**Cannot start AcmePro** If you receive the error message when trying to login, make sure the management server is running, and that the AcmePro server processes are running on the management server.

Describing problems

When describing a problem, focus on the problem, not the cause of the problem.

Change	To
Problem If you receive an error message when trying to login, the AcmePro server processes may not be running on the management server.	**Problem** After you login, the GUI fails to open. You receive the following error message: `Management server is not running.`

TIP For related guidelines, see "Topics" on page 140.

Explaining solutions

When explaining solutions, use step-by-step procedures. Format procedures as ordered lists. Format options as itemized lists.

<table>
<tr><th>Change</th><th>To</th></tr>
<tr><td>Solution

Determine the process ID of the endlessly running action using the <code>ps</code> command. Issue a <code>kill</code> command for the specific process ID. To prevent the problem in the future, the related template condition should be changed, or the script or command should be fixed.</td><td>Solution

Do the following:

1 Enter the <code>ps</code> command to determine the process ID of the endlessly running action.

2 Enter a <code>kill</code> command for the specific process ID.

3 To prevent the problem from recurring, do one of the following:
• Change the related template condition.
• Fix the script or command.</td></tr>
<tr><td><b>Solution</b>

Try the following:

▪ Make sure that the management server is running. Try the <code>ping</code> command from an MS-DOS terminal window:
<code><b>ping <<i>management_server</i>></b></code>

▪ The processes may not be running on the management server. Ask your administrator to check the status of the server processes.</td><td><b>Solution</b>

Do the following:

1 Make sure the management server is running.
From an MS-DOS terminal window, enter the <code>ping</code> command:
<code><b>ping <<i>management_server</i>></b></code>

2 Make sure the server processes are running on the management server.
Ask your administrator to check the status of the server processes.</td></tr>
</table>

TIP To related guidelines, see the following sections:

- "Itemized lists" on page 98
- "Procedures" on page 118

Grouping related problems and solutions

Group closely related problems and solutions into one troubleshooting scenario. Separate each distinct subproblem and subsolution within the scenario. Use a consistent subheading syntax to map subproblems to subsolutions (for example, "Problem *X*" and "Solution *X*").

Change	To
Application startup problem An application can no longer be started on a managed node. **Description** ■ An application has been upgraded, and its command path has been changed. ■ User's password for default application startup has been changed. **Solution** ■ Adapt the default application startup accordingly. ■ Use customized application startup options. See "To customize the startup attributes of an application" on page 31 for more information.	**Cannot start an application** If you can no longer start an application on a managed node, adapt the default application startup configuration. Or use customized startup options. **Problem A** An application has been upgraded, and its command path has been changed. **Solution A** Adapt the default application startup accordingly. **Problem B** The user password for the default application startup has been changed. **Solution B** Use customized application startup options. For details, see "To customize the startup attributes of an application" on page 31.

CHAPTER 4

Configuring language

You can "configure" language to increase the usability of your documents. If your language "defaults" are user-centered, your writing results in documents that are easy to use. For example, if you write in active voice, second person, and present tense, you speak directly to your users. Likewise, if you use parallel construction in headings, captions, list items, and so on, you make it easier for users to scan documents for the information they need. Finally, if you write concise sentences, you provide users with information chunks they can understand without thinking.

You can also configure language to increase the re-usability of your documents. If you develop shared modules that follow a default writing style, you can assemble the modules into many different documents seamlessly. Although inconsistent use of abbreviations, capitalization, and punctuation is not fatal to document usability, it is unprofessional. You can use language defaults to present users with documents that appear to be written by one author only.

To supplement your in-house style guide, this chapter presents language guidelines that help improve the usability and re-usability of single-source documents. Not every guideline fits every project. Choose the guidelines you need. Then modify them to fit your corporate and project requirements.

In this chapter

This chapter contains the following language guidelines:

Abbreviations 151

To enable users to identify abbreviations anywhere, spell them out the first time they are mentioned in a module. Use common and trademarked abbreviations. Avoid uncommon abbreviations.

Capitalization 154

Follow a consistent capitalization scheme for document titles, captions, commands, definition lists, filenames, glossaries, and itemized lists.

Parallel construction 158

Use the same syntax structure for elements with the same function. Use parallel construction in captions, glossaries, headings, indexes, introductions, lists, steps, sentences, and tables.

Person 167

Address your primary audience in second person singular. Address your secondary audience in third person plural.

Punctuation 169

Use punctuation to make your writing not just grammatically correct, but easy to understand in any context.

Sentence construction 171

To prepare modules for online documents, shorten and simplify sentences. If a sentence contains many serial items, convert the items into an itemized list.

Tense 173

"Now" is wherever users are at the moment. To build user-centered modules, use present tense wherever possible.

Voice 177

To speak clearly and directly to users, write sentences and phrases in active voice rather than passive voice.

Abbreviations

Abbreviations are shortened forms of words and phrases. To enable users to easily identify abbreviations anywhere, spell them out the first time they are mentioned in a module. Use common and trademarked abbreviations. Avoid uncommon abbreviations, especially Latin abbreviations.

Types of abbreviations

There are three types of abbreviations:

- **Common abbreviations**
  Common abbreviations (for example, "HTML") are usually more familiar to users than the words they abbreviate (for example, "HyperText Markup Language"). For this reason, you should *not* spell out common abbreviations after their first mention in a module.
- **Trademarked abbreviations**
  Trademarked abbreviations (for example, "IBM") are legally shortened names for products or companies (for example, "International Business Machines, Inc."). These proprietary abbreviations are usually more familiar to users than the words they abbreviate. For this reason, you should *not* spell out trademarked abbreviations after their first mention in a module.
- **Uncommon abbreviations**
  Uncommon abbreviations (for example, "UMB") are usually less familiar to users than the words they abbreviate (for example, "upper memory block"). For this reason, you should spell out uncommon abbreviations wherever you use them.

TIP Whether an abbreviation is common or uncommon depends on your target audience. If you are developing single-source documents for more than one audience, favor the less sophisticated audience. If you are not sure whether all of your users are familiar with a given abbreviation, spell it out.

Common abbreviations

To make common abbreviations understandable, even when modules are re-used, spell out the abbreviations the first time you mention them in a primary module. Thereafter, use the abbreviations without spellouts.

First mention	Thereafter
Extensible Markup Language (XML)	XML
Portable Document Format (PDF)	PDF

Latin abbreviations

Latin abbreviations are often unfamiliar to users. Use English equivalents of Latin abbreviations.

Change	To
e.g.,	for example,
etc.	and so on
i.e.,	that is,

TIP The abbreviation for the Latin word "versus" ("v.") is commonly understood. This abbreviation is useful in indexes, where space is extremely limited.

Trademarked abbreviations

Trademarked abbreviations are usually more familiar to users than the words they abbreviate. Use trademarked abbreviations rather than the words they abbreviate.

Change	To
International Business Machines	IBM
Institute of Electrical and Electronics Engineers	IEEE

Uncommon abbreviations

Avoid abbreviations that are not familiar to your users. Spell out uncommon abbreviations wherever they appear.

Change	To
CRC	Cyclical Redundancy Checking
DES	Data Encryption Standard
UMB	upper memory block

Uncommon words

Occasionally, not only are abbreviations more familiar to users than the words they abbreviate, but the words they abbreviate are all but unknown to users. In these instances, present the abbreviations with their spellouts in parentheses the first time you mention them in a primary module. Thereafter, use the abbreviations without spellouts.

First mention	Thereafter
TIFF (Tagged Image File Format)	TIFF
GIF (Graphics Interchange Format)	GIF
JPEG (Joint Photographic Experts Group)	JPEG

Capitalization

Consistent capitalization improves document usability. Follow consistent capitalization in document titles, captions, commands, definition lists, filenames, glossaries, and itemized lists.

Types of capitalization

There are four ways you can capitalize words:

- **Uppercase (ABC)**

  Uppercase is normally used for literal strings, such as computer commands and filenames (for example, "`AUTOEXEC.BAT`").

- **Lowercase (abc)**

  Lowercase is normally used for glossary terms (for example, "database").

- **Initial capitalization (Abc Def)**

  Initial (title) capitalization is normally used for product elements, such as computer program names and menu items (for example, "Print Preview").

- **Downstyle capitalization (Abc def)**

  Downstyle (sentence) capitalization is normally used for captions, headings, and lists (for example, "About this guide").

NOTE Some publishing organizations prefer initial capitalization to downstyle capitalization for captions and headings, particularly in online help. Whichever form of capitalization you choose, make sure to follow it consistently in headings and captions.

Capitalizing captions

For figure and table captions, use downstyle capitalization (Abc def) rather than initial capitalization (Abc Def). Initial capitalization makes it difficult for users to distinguish between proper and non-proper nouns.

Change	To
Figure 1-2 Print *Dialog Box*	Figure 1-2 Print *dialog box*

TIP For related guidelines, see "Captions" on page 51.

Capitalizing commands

Commands and their parameters are case-sensitive in some operating systems. When listing commands, follow the capitalization rules of the operating system on which your product is based.

DOS	UNIX
`DIR /P`	`chmod 705 private`

Capitalizing definition lists

In definition lists, some commands, parameters, and so on are case-sensitive. For terms, follow the capitalization rules of the elements you are defining. For definitions, use downstyle capitalization (Abc def).

Change	To
`MESSAGE KEY`	`Message Key`
Key Associated with the Message.	Key associated with the message.
`message no.`	`Message No.`
unique identification number associated with the message. This number enables you to program with open APIs.	Unique identification number associated with the message. This number enables you to program with open APIs.

TIP For related guidelines, see "Definition lists" on page 63.

Capitalizing document titles

Initial capitalization (Abc Def) is an industry standard for document titles. When capitalizing document titles, follow this industry standard.

Change	To
AcmePro for UNIX installation guide	AcmePro for UNIX Installation Guide

Capitalizing filenames

Filenames are case-sensitive in some operating systems. When listing filenames, follow the capitalization rules of the operating system on which your product is based.

DOS	UNIX
`AUTOEXEC.BAT`	`/cgi-bin/email.cgi`

Capitalizing glossaries

For glossary terms that are not proper nouns, use lowercase (abc). Using uppercase (ABC DEF), initial capitalization (Abc Def), or downstyle capitalization (Abc def) adds visual "noise" to glossary terms.

Change	To
Modular writing Element-based writing method.	**modular writing** Element-based writing method.
SGML standard generalized markup language. Generic markup language used to represent documents.	SGML Standard Generalized Markup Language. Generic markup language used to represent documents.

TIP For related guidelines, see "Glossaries" on page 81.

Capitalizing headings

For headings, use downstyle capitalization (Abc def). Initial capitalization (Abc Def) makes it difficult for users to recognize proper nouns.

Change	**To**
Installing an Oracle *Database*	Installing an Oracle *database*

TIP For related guidelines, see "Headings" on page 85.

Capitalizing itemized lists

For itemized lists, use downstyle capitalization (Abc def). Lowercase (abc def) makes it difficult for users to see patterns (for example, alphabetical ordering) when scanning list items. Initial capitalization (Abc Def) makes it difficult for users to distinguish between proper and non-proper nouns.

Change	**To**
This section describes the following main menus:	This section describes the following main menus:
▪ File *Menu*	▪ File *menu*
▪ Edit *Menu*	▪ Edit *menu*
▪ View *Menu*	▪ View *menu*

TIP For related guidelines, see "Itemized lists" on page 98.

Parallel construction

Parallel construction is using the same phrase or sentence structure for elements with the same function. By using the same structure, you express the equality of ideas between similar elements. Use parallel construction in captions, glossaries, indexes, introductions, lists, procedures, processes, sentences, and tables.

Parallel captions

Use parallel construction for figure and table titles. In figure captions, describe content from the system perspective or from the user perspective. In table captions, describe content from the system perspective only.

Change		To	
Figure 2-1	Reviewing the nodes in a node group	Figure 2-1	Nodes in a node group
Figure 2-2	New message group	Figure 2-2	New message group
		Figure 2-1	Reviewing the nodes in a node group
		Figure 2-2	Adding a new message group
Table 3-1	Backing up data	Table 3-1	Backup methods
Table 3-2	Event and message correlation	Table 3-2	Event and message correlation

TIP For related guidelines, see the following sections:

- "Captions" on page 51
- "Figures" on page 74
- "Tables" on page 129

Parallel glossary entries

In glossary entries, use parallel construction for terms, definitions, and cross-references. For example, use singular nouns for terms, and begin definitions with nouns. Likewise, order "see" and "see also" references alphabetically. Separate each reference with semicolons (;). To emphasize parallel construction in definitions, repeat "helper" words (for example, articles, conjunctions, prepositions, pronouns, and verbs) in each clause.

Change	To
control agents Agents on each managed node that are responsible for starting and stopping all other agents, and processing requests from the management server. Control agents are sometimes called "opcctla." For related entries, see control manager, control switch.	**control agent** Also known as "opcctla." Agent on each managed node that starts and stops all other agents. This agent also processes requests from the management server. *See also* control manager; control switch.
control manager Also known as *opcctlm*. A process on the management server that starts and stops all other manager processes. The control manager checks that all manager processes are running. *See also* control switch and control agents.	**control manager** Also known as "opcctlm." Process on the management server that starts and stops all other manager processes. This process also verifies that all manager processes are running. *See also* control agent; control switch.
control switch To switch the responsibility of a message from source to target management servers.	**control switch** Process of switching the responsibility for a message from the source management server to target management servers. *See also* control agent; control manager.

TIP For related guidelines, see "Glossaries" on page 81.

Parallel headings

In headings, use parallel construction to identify distinct types of modules.

Module type	Heading syntax
Definition lists	Types of dialog boxes
Definition list 1	About the ABC dialog box
Definition list 2	About the DEF dialog box
Itemized lists	Installation requirements
Itemized list 1	Hardware requirements
Itemized list 2	Software requirements
Procedures	Installing AcmePro
Procedure 1	To install AcmePro automatically
Procedure 2	To install AcmePro manually
Processes	Problem solving
Process 1	Detecting problems
Process 2	Investigating problems
Process 3	Solving problems
Process 4	Documenting solutions
Topics	About Acme Pro
Topic 1	Who should use AcmePro
Topic 2	What AcmePro does
Troubleshooting scenarios	Troubleshooting runtime problems
Scenario 1	Cannot start AcmePro
Scenario 2	Cannot start an application
Scenario 3	Cannot execute commands

TIP For related guidelines, see "Headings" on page 85.

Parallel index entries

Wherever possible, use parallel construction for same-level index entries. For example, begin a primary entry with a gerund (that is, a verb ending in "ing"), and begin its secondary entries with nouns.

Change	To

correlate
    event, 48
    messages, 241
creating
    configuration files, 252–256
    fast link, 75
    message source templates, 197
	correlating
	    events, 48
	    messages, 241
	creating
	    configuration file, 252–256
	    fast link, 75
	    message source template, 197

TIP For related guidelines, see "Indexes" on page 92.

Parallel introductions

Use parallel construction when introducing definition lists, itemized lists, figures, procedures, processes, and tables. Aim the introduction in the direction of the module you are introducing.

Change	To
The following options are contained in the View menu:	The View menu contains the following options:
Your Solaris managed nodes should meet the following hardware requirements before you install AcmePro:	Before installing AcmePro, make sure your Solaris managed nodes meet the following hardware requirements:
Perform the following tasks in the specified order to configure AcmePro successfully:	To configure AcmePro, follow these steps:

Parallel list items

In definition lists, use parallel construction for terms and definitions. For example, you could use singular nouns for terms. And you could begin all definitions with action verbs. In itemized lists, use parallel construction for same-level list items. For example, you could begin list "headings" with gerunds (that is, verbs ending in "ing"). And you could begin annotations with action verbs.

Change	To

`Expand Object Pane`
When you press **Shift+F10+E**, all branches of the object tree expand. Equivalent to clicking + next to all branches of the object tree.
`Collapse Object Pane`
This is the equivalent of clicking the minus sign (-) next to all branches of the object tree. This option collapses all branches of the object tree. To collapse the object pane, press **Shift+F10+C**.
	`Expand Object Pane` (**Shift+F10+E**)
	Expands all branches of the object tree. Equivalent to clicking the plus sign (+) next to all branches of the object tree.
	`Collapse Object Pane` (**Shift+F10+C**)
	Collapses all branches of the object tree. Equivalent to clicking the minus sign (-) next to all branches of the object tree.

AcmePro has the following benefits: ▪ You reduce the time lost by operators as a result of system downtime. ▪ Preventive actions help you prevent problems. ▪ Cost reductions in managing your client-server environment	AcmePro helps you do the following: ▪ **Reduce downtime** Reduce the time lost by operators as a result of system downtime. ▪ **Reduce problems** Reduce the number of problems through preventive actions. ▪ **Reduce costs** Reduce the cost of managing your client-server environment.

TIP For related guidelines, see the following sections:

- "Definition lists" on page 63
- "Itemized lists" on page 98

Parallel procedure steps

Use parallel construction for all procedure steps. For example, format each distinct user action as a step. Include details in a separate paragraph.

Change	To
To modify the attributes of a message, follow these steps: 1 First, select the message you want to modify in the message browser. 2 The Modify Message Attributes dialog box opens when you right-click the message, then select `Modify` from the popup menu. 3 Modify the message severity or text in the Modify Message Attributes dialog box. 4 When you click `[OK]`, your changes are applied immediately to the message you selected. You now own the message.	To modify the attributes of a message, follow these steps: 1 In the message browser, select the message you want to modify. 2 Right-click the message, then select `Modify` from the popup menu. The Modify Message Attributes dialog box opens. 3 In the Modify Message Attributes dialog box, change one of the following: • Severity • Text 4 Click `[OK]`. Your changes are applied immediately to the message you selected. You now own the message.

TIP For related guidelines, see "Procedures" on page 118.

Parallel process steps

Use parallel construction for steps in processes. For example, begin each step with a noun or a gerund (that is, a verb ending in "ing").

Change	To
The sequential indexing method includes 10 steps:	The sequential indexing method includes 10 steps:
1 Chapter *indexing*	1 *Indexing* chapters
2 Procedure *indexing*	2 *Indexing* procedures
3 Topic *indexing*	3 *Indexing* topics
4 *Index* product names	4 *Indexing* product names
5 *Index* product components	5 *Indexing* product components
6 *Index* front and back matter	6 *Indexing* front and back matter
7 About *editing*	7 *Editing* your index
8 "See" reference *creation*	8 *Creating* "see" references
9 *Creating* "see also" references	9 *Creating* "see also" references
10 Index *testing*	10 *Testing* your index

TIP To find out how to build processes, see "Processes" on page 124.

Parallel sentences

In sentences, use parallel construction for words or phrases separated by commas or semicolons. To emphasize parallel construction in sentences, repeat "helper" words (for example, articles, conjunctions, prepositions, pronouns, and verbs) in each clause. To test parallel constructions in sentences, convert them into itemized lists.

Change	To
As an operator, *you can* save console session settings and reload assigned defaults.	As an operator, *you can* save console session settings, and *you can* reload assigned defaults.
	As an operator, *you can* do the following: ▪ Save console session settings. ▪ Reload assigned defaults.
In the basic and advanced search *options*, the search begins at the top of the object tree, rather than from the selected item.	*In the* basic search *option* and *in the* advanced search *option*, the search begins at the top of the object tree, rather than from the selected item.
	Searching begins at the top of the object tree, rather than from the selected item, *in both search options*: ▪ Basic search ▪ Advanced search

TIP For related guidelines, see the following sections:

- "Itemized lists" on page 98
- "Sentence construction" on page 171

Parallel table cells

Use parallel construction for headings and cells. For example, begin all phrases in parallel cells with a noun or a verb.

Change

Table 3-1 Responsibilities of *system and template administrators*

Task	Responsible Administrator
Message policy *definition*	System Administrator
Implementing the message policy	Template Administrators
Template group *configuration*	Template Administrators
Assigning templates and groups	Template Administrators
Distribute templates and groups	System Administrator

To

Table 3-1 Responsibilities of *the system administrator and the template administrators*

	Responsible Administrator	
Task	System	Template
Define the message policy	✓	
Implement the message policy		✓
Configure template groups		✓
Assign templates and template groups		✓
Distribute templates and template groups	✓	

TIP For related guidelines, see "Tables" on page 129.

Person

Person is the form of a personal pronoun that indicates whether the pronoun represents the speaker (first person), the person spoken to (second person), or the person or thing spoken about (third person). Address your primary audience in second person singular. Address your secondary audience in third person plural.

TIP For related guidelines, see the following sections:

- "Tense" on page 173
- "Voice" on page 177

Types of person

There are three forms of personal pronoun:

- **First person**

  First person is almost never used in product documentation because it refers to writers rather than to users.

  Singular: I

  Plural: we

- **Second person**

  Second person enables you to speak directly to users. It also enables you to use imperatives in procedures (for example, "press **Enter**"). Normally, second person is reserved for primary users only.

  Singular: you

  Plural: you

- **Third person**

  Third person places a third party between you and your users. Normally, third person is reserved for secondary users only (for example, when referring to operators in a document written for administrators).

  Singular: he | she | it

  Plural: they

Of the three forms, second person is the most direct. For this reason, second person should be your default for primary users.

Types of audiences

If you have more than one type of user, do the following:

- **Primary audience**

  Clearly equate second person (for example, "you") with your primary audience (for example, system administrators).

- **Secondary audience**

  Clearly equate the appropriate title (for example, "operators") with your secondary audience (for example, system operators).

Second person for primary users

Address primary users in second person singular (for example, "you"), not third person plural (for example, "operators").

Change	To
Operators currently using a version of AcmePro prior to 5.0 must upgrade to version 5.0 before *they* can upgrade to version 6.0.	If *you* are currently using a version of AcmePro lower than 5.0, *you* must upgrade to version 5.0 before *you* can upgrade to version 6.0.

Third person for secondary users

Address secondary users in third person plural (for example, "operators"), not second person singular (for example, "you").

Novice as secondary user	Expert as secondary user
Administrators can use runtime data and historical data to generate reports. Historical data can also be helpful when *administrators* create instructions to help *you* solve problems caused by similar events.	*You* can use runtime data and historical data to generate reports. Historical data can also be helpful when *you* create instructions to help *operators* solve problems caused by similar events.

TIP To avoid awkward gender constructions, use plural for secondary audiences (for example, use "operators" instead of "the operator").

Punctuation

Use commas, hyphens, parentheses, and periods to break texts into small chunks that are easy to understand online. Avoid colons and semicolons.

Colons

Colons (:) increase sentence length, and decrease usability, especially online. To shorten sentences, use periods instead of colons.

Change	To
You need to configure the AcmePro on only one of the cluster nodes: all necessary configuration files are copied to the other cluster nodes during configuration.	You need to configure the AcmePro package on only one of the cluster nodes. All necessary configuration files are copied to the other cluster nodes during configuration.

TIP Although colons increase sentence length, they are ideal for punctuating introductions to definition lists, itemized lists, and ordered lists.

For related guidelines, see the following sections:

- "Introducing definition lists" on page 64
- "Introducing itemized lists" on page 99
- "Introducing procedures" on page 119

Commas

Commas break sentences into chunks that can be understood easily. To shorten sentences, use as many commas as are grammatically allowed.

Change	To
AcmePro detects, solves and prevents problems that occur in networks, systems and applications within your IT environment.	AcmePro detects, solves, and prevents problems that occur in networks, systems, and applications within your IT environment.

TIP For related guidelines, see "Formatting sentences as lists" on page 172.

Parentheses

Parentheses provide a clear visual distinction between statements and examples of those statements. In sentences, separate simple examples with parentheses rather than with commas.

Change	To
Product names are generally listed according to their respective product numbers, for example, AP5000.	Product names are generally listed according to their respective product numbers (for example, AP5000).

TIP Do not place long, complex examples within parentheses. For details, see "Examples" on page 67.

Semicolons

Semicolons (;) increase sentence length, and decrease usability, especially online. To shorten sentences, use periods instead of semicolons.

Change	To
The subdirectory prefix `/u01` is the recommended default; you can use other appropriate prefixes if needed.	Use the subdirectory prefix `/u01` as your default. You can change this default, if needed.

TIP For related guidelines, see "Sentence construction" on page 171.

Sentence construction

To build usable and re-usable texts, shorten and simplify sentences. If a sentence contains serial items, format the items as an itemized list.

Simplifying sentences

Simplify sentence construction to increase usability. To simplify sentences, use active voice and present tense.

Change	To
The advantage of suppressing duplicate messages on the server is that high system loads caused by large numbers of messages can be significantly reduced.	By suppressing duplicate messages on the server, you significantly reduce the number of messages. By reducing the number of messages, you significantly reduce the load on the server.

TIP For related guidelines, see the following sections:

- "Tense" on page 173
- "Voice" on page 177

Shortening sentences

Divide long sentences into shorter sentences to increase usability. Divide long serial phrases into separate sentences. If necessary, begin some of these sentences with "And," "Or," or "Then." For serial phrases, use parallel construction.

Change	To
You can assign templates to nodes or node groups, or assign template groups to nodes or node groups where the message interception should be performed, then you can distribute the new configuration.	You can assign templates to nodes or node groups. Or you can assign template groups to nodes or node groups where the message interception should be performed. Then you can distribute the new configuration.

TIP For related guidelines, see "Parallel construction" on page 158.

Formatting sentences as lists

If a sentence contains many serial items, format the serial items as itemized list items. Use parallel construction for the list items.

Change	To
You can view database reports in a window, save them to file, and print them.	With database reports, you can do the following: ▪ Display reports in a window ▪ Save reports to file ▪ Print reports

TIP For related guidelines, see the following sections:

- "Itemized lists" on page 98
- "Parallel list items" on page 162
- "Parallel sentences" on page 165

Tense

Users view documentation in real time. "Now" is wherever users are at the moment. To maintain this user-centered perspective, use present tense wherever possible. Use past or future tense only when referring to something that occurs before or after the current phrase or sentence.

TIP For related guidelines, see the following sections:

- "Person" on page 167
- "Voice" on page 177

About user-centered time

Present tense tells users "you are here now":

- **Paragraphs**

  Even though the first paragraph of a topic is located before the second paragraph of the topic, both are timeless. When users read the first paragraph, they associate "now" with the first paragraph. When users move to the second paragraph, "now" moves with them to the second paragraph.

- **Steps**

  Even though the first step of a procedure or process is located before the second step of the procedure or process, both steps are timeless. When users read the first step, they associate "now" with the first step. When users move to the second step, "now" moves with them to the second step.

Occasionally, you need to refer to something that occurs before or after the current phrase or sentence. In these instances, make sure that a change in tense is absolutely necessary. If past or future tense is not absolutely necessary, use present tense.

Change	To
After you *have set up* your printer, you can print a test document.	After you *set up* your printer, you can print a test document.

Present tense in procedures

To keep procedures user-centered, write each step of every procedure in present tense. Use past or future tense only when referring to something that occurs before or after the current phrase or sentence.

Change	To
To view a message node automatically	**To view a message node automatically**
After *detecting* a problem, you *will be able to* automatically highlight the affected node in the object pane.	Once you *detect* a problem, you *can* automatically highlight the affected node in the object pane.
To access a message node automatically, *you will need to follow* these steps:	To access a message node automatically, *follow* these steps:
1 In the message browser, *you will select* the message you investigated.	1 In the message browser, *select* the message you investigated.
2 Right-click the message. A popup menu *will appear*.	2 Right-click the message. A popup menu *appears*.
3 In the popup menu, select `Node in Object Pane`. This option *will highlight* the node or nodes in the object pane. You *will then be able to start* applications on that node.	3 In the popup menu, select `Node in Object Pane`. This option *highlights* the node or nodes in the object pane. You *can now start* applications on that node.

TIP For related guidelines, see "Procedures" on page 118.

Present tense in processes

To keep processes user-centered, write each step of every process in present tense. Use past or future tense only when referring to something that occurs before or after the current phrase or sentence.

Change	To
About the problem solving process	**About the problem solving process**
The AcmePro problem solving process includes the following steps:	The AcmePro problem solving process includes the following steps:
1 **Identify the problem.** An error message *will indicate* where and why the problem occurred. The message *will suggest* what you should do to resolve the problem.	1 **Identify the problem.** An error message *indicates* where and why the problem occurs. The message *suggests* what you should do to resolve the problem.
2 **Solve the problem.** To solve the problem, you *will use* the following tools: • Automatic actions • Broadcast commands • Operator-initiated actions	2 **Solve the problem.** To solve the problem, you *can use* the following tools: • Automatic actions • Broadcast commands • Operator-initiated actions
3 **Document the solution.** To record how you *solved* the problem, you *will add* a message annotation.	3 **Document the solution.** To record how you *solved* the problem, you *add* a message annotation.
4 **Close the problem.** You *will close* the original error message to move it from the current work area to the history database.	4 **Close the problem.** You *close* the original error message to move it from the current work area to the history database.

TIP For related guidelines, see "Processes" on page 124.

Present tense in topics

To keep topics user-centered, write each sentence of every topic in present tense. Use past or future tense only when referring to something that occurs before or after the current phrase or sentence.

Change	To
When the search function finds a specific item, the item *will be highlighted* in the object pane. If the item *was not visible* because the object tree *was collapsed*, the tree *will be expanded*, and the item *will be scrolled* into the visible area of object pane. Normally, the search *will start* from the top of the object tree. However, if you *started* a search while an item *was highlighted*, the search *will start* from the selected item.	When the search function finds a specific item, the item *is highlighted* in the object pane. If the item *is not visible* because the object tree *is collapsed*, the tree *is expanded*, and the item *is scrolled* into the visible area of object pane. Normally, the search *starts* from the top of the object tree. However, if you *start* a search while an item *is already highlighted*, the search *starts* from the selected item.

TIP For related guidelines, see "Topics" on page 140.

Voice

To speak clearly and directly to users, write phrases and sentences in active voice rather than passive voice.

TIP For related guidelines, see the following sections:

- "Person" on page 167
- "Tense" on page 173

Types of voice

There are two types of voice:

- **Active voice (subject, verb, object)**
  Makes phrases and sentences *more* direct and forceful.
- **Passive voice (object, verb, subject)**
  Makes phrases and sentences *less* direct and forceful.

Whenever possible, use active voice rather than passive voice.

TIP In any given phrase or sentence, the easiest way to correct passive voice is to determine who or what is performing the action. Then make that person or thing the subject, rather than the object, of the sentence.

Active voice in definition lists

In definition list items, use active voice whenever possible.

Change	To
`Print` With this option, messages can be printed from the message browser.	`Print` Enables you to print messages from the message browser.

TIP For related guidelines, see "Definition lists" on page 63.

Active voice in itemized lists

In itemized lists, use active voice whenever possible.

Change	To
The following can be done with database reports: ▪ Window display ▪ File saving ▪ Document printing	With database reports, you can do the following: ▪ Display reports in a window ▪ Save reports to file ▪ Print reports

TIP For related guidelines, see "Itemized lists" on page 98.

Active voice in procedures

In procedure steps, use active voice and imperatives (that is, commands with an implied "you") whenever possible.

Change	To
To print a document, follow these steps: 1 `Print` should be selected from the File menu. The Print dialog box appears. 2 `[OK]` is clicked in the Print dialog box.	To print a document, follow these steps: 1 From the File menu, select `Print`. The Print dialog box appears. 2 In the Print dialog box, click `[OK]`.

TIP For related guidelines, see "Procedures" on page 118.

Active voice in processes

In processes, use active voice whenever possible.

Change	To
Problem solving There are three steps in the problem solving process. First, you are notified exactly when and where a problem occurs, including where and why the problem occurred, and suggestions what you should do to resolve the problem....	**Problem solving** The problem solving process includes the following steps: 1 **Detecting problems** When a problem occurs, an error messages notifies you. The message explains exactly where, when, and why the problem occurred. And it suggests steps you can take to solve the problem. 2 ...

TIP For related guidelines, see "Processes" on page 124.

Active voice in topics

In topics, use active voice whenever possible.

Change	To
Duplicate message *suppression* works on the following principle: the event, for example a logfile entry, is compared with a condition.	AcmePro *suppresses* duplicate messages by comparing an event (for example, a logfile entry) with a condition.

TIP For related guidelines, see "Topics" on page 140.

CHAPTER 5

Leveraging technology

You can leverage technology to automate many parts of your single sourcing process. For example, you can use conditional text to control which content modules of your single-source documents appear in which formats, languages, and versions of your output documents. Likewise, you can use variables for text you expect to change (for example, product names or versions).

This chapter provides you with an overview of authoring tools, conversion tools, and content management systems (CMS) used to build single-source documents. It also provides guidelines to using important functions of these tools (for example, conditional text, search engines, and variable text). Finally, this chapter shows you how to leverage these functions to save time and money when building complex documents for multilingual audiences.

The guidelines in this chapter are organized alphabetically, for easy reference. Each guideline is cross-referenced to related guidelines. Although each guideline has proven itself in successful single sourcing projects, not every guideline is suitable for every project. Choose the guidelines you need. Then modify them to fit your projects.

In this chapter

This chapter contains the following guidelines:

Conditional text 183

Conditional text enables you to mark elements for conditional display, then switch these elements on or off. Use conditional text for specific document types, formats, languages, and versions.

Conventions 186

To build document conventions that work anywhere, do not hard-code character formats. If possible, use element tags. Otherwise, define character styles that can be converted to other formats.

Development tools 188

Single sourcing tools are document-driven or database-driven. The best development tools are based on SGML or XML, both of which separate content from format.

Filenames 192

Intuitive filenames are important in large single sourcing projects. Input filenames should indicate module content and type. Output filenames should indicate module content, type, and parentage.

Localization 195

Localization involves document translation and customization for local markets. Use conditional text for customized elements. To save time and money, translate modules before assembly.

Search engines 198

Search engines enable users to search documents by title, header, text, and index. To maximize the effectiveness of search engines, integrate search and index functions.

Variables 200

Variables are definitions for dynamic text. Create variables when you expect text to change. Define content variables in authoring tools. Define format variables in conversion tools.

Conditional text

Conditional text is a function in many authoring and conversion tools. This function enables you to mark text or graphics for conditional display, then switch the sections on or off, as needed. For example, you can mark some sections of a document for online help only, then switch them off before printing the document. Use conditional text for specific document types, formats, languages, and versions. Avoid conditional text if it causes duplicate work.

When to use conditional text

You can use conditional text to control the display of modular content:

- **Types**
  Mark text for different document types (for example, an installation guide, a user guide, and a reference manual).
- **Formats**
  Mark text for different document formats (for example, a printed manual, an online help system, and a training workbook).
- **Languages**
  Mark text for different languages (for example, English and Japanese).
- **Versions**
  Mark text for different versions of the same document (for example, a standard version and an enterprise version).

Use conditional text for modules associated with different document types, formats, languages, and versions. Use unconditional text for all shared modules.

TIP For related guidelines, see the following sections:

- "About authoring tools" on page 189
- "About conversion tools" on page 190
- "Customizing modular content" on page 197

When not to use conditional text

Do not use conditional text if it causes duplicate work. For example, do not use conditional text to solve conflicts between two document indexes with two different formats.

Avoid duplicate content tags

If you are using a single source to develop two different documents with two incompatible index structures, you might be tempted to use conditional text to create two sets of index tags, one for each document. Such use of conditional text would cause duplicate work. To avoid duplicate work, you could create a master index shared by the two documents. This master index would enable you to create one set of index tags without using conditional text at all.

Index types	Index entries
Index A	printing documents, 51
Index B	printing documents to file, 52
Master index	printing documents to file, B-52 to printer, A-51

Avoid duplicate format tags

If you are using a single source to develop indexes for one print document and one online document, you might be tempted to use conditional text to create two sets of index tags, one for each format. Such use of conditional text would cause duplicate work. To avoid duplicate work, you could create one set of index tags to generate a single-source index in one format, then convert the index to the other format automatically.

TIP For related guidelines, see the following sections:

- "About conversion tools" on page 190
- "Building master indexes" on page 95
- "Formatting index entries" on page 94

Formatting conditional text

You can format conditional text so that it is easy to see in your authoring tool. Depending on your authoring tool, you may be able use colors or character styles to identify different conditional text tags. However you format your conditional text, make sure the tags do not interfere with output formats. For example, if you use blue for conditional text used in printed manuals, the manuals may not print properly.

Document	Tag	Color	Style
Format	`Print`	Black	Plain
	`Online`	Blue	**Bold**
Language	`English`	Red	Underline
	`Japanese`	Orange	Double underline
Type	`InstallGuide`	Green	***Bold italics***
	`UserGuide`	Yellow	*Italics*
Version	`Public`	Purple	Overline
	`Archive`	Gray	~~Strikethrough~~

Combining conditional text

You can combine conditional text tags for a wide variety of purposes. If you find yourself using 10 or more conditional text tags in your single-source documents, consider using a content management system (CMS).

Document A	Document B	Document C	Document D
`Print`	`Online`	`Print`	`Online`
`English`	`English`	`Japanese`	`Japanese`

TIP For a list of standard functions provided by content management systems, see "About content management systems" on page 191.

Conventions

Conventions are the typographical rules you follow throughout a given document. To build flexible conventions that are not tied to a particular document format, do not hard-code character formats. With content-based authoring tools, set conventions with element tags. With format-based authoring tools, set conventions with character styles.

Types of conventions

You can set document conventions in one of two ways:

- **Content-based conventions**

  With content-based authoring tools, you define dynamic conventions with element tags. You can then use conversion tools to convert the conventions to other document formats.

  For details, see "Defining conventions with element tags" on page 187.

- **Format-based conventions**

  With format-based authoring tools, you define static conventions at the character level. To make it possible for conversion tools to convert the conventions to other document formats, you need to first define and use character styles in your authoring tool.

  For details, see "Defining conventions with character styles" on page 187.

Dynamic, content-based conventions are ideal for single sourcing. Not only do they enable you to convert documents automatically, they also enable you to correct tagging mistakes in your single-source documents long before conversion begins. And they enable you to focus your attention on content rather than format. Content-based authoring tools take the drudgery out of document conventions.

TIP Listed conventions benefit writers more than users. These lists rarely appear in online documents, and are rarely read in print documents. The best place to list conventions is in a project style guide. By listing conventions in a project style guide, you help writers to follow single-sourcing conventions in shared modules, thereby eliminating the need for rework.

Defining conventions with element tags

If you are using a content-based authoring tool, you can define dynamic document conventions with element tags. For example, if you are using Adobe FrameMaker+SGML, you might mark system messages with the DocBook element `ComputerOutput`. Likewise, you might mark user input with the DocBook element `UserInput`. If you wanted, you might then use Quadralay WebWorks Publisher to convert these elements into different formats for online documents.

If you are using a content-based authoring tool, format characters with element tags, not character tags.

Element style	Character style	Sample output
`ComputerOutput`	`Monospace`	`File not found`
`UserInput`	`MonospaceBold`	**`chmod 705 private`**

Defining conventions with character styles

If you are using a format-based authoring tools, you can define document conventions in character styles. Content (meaning) is tied to format (typography). For example, if you are using Adobe FrameMaker, you might define a monospaced character style for system messages. Likewise, you might define a bolded monospaced character style for user input. If you wanted, you could then use Quadralay WebWorks Publisher to convert these character styles into different online formats.

If you are using a format-based authoring tool, format characters with character styles, not hard-coded character formats.

Character style	Character format	Sample output
`Monospace`	Courier, Plain	`File not found`
`MonospaceBold`	Courier, Bold	**`chmod 705 private`**

TIP For related guidelines, see "Development tools" on page 188.

Development tools

Although there are many different single sourcing tools on the market, most fall under one of three categories:

- **Authoring tools**

  Authoring tools are used to develop documents. Typically, they are desktop publishing applications you can use to develop multimedia documents from a single source (for example, Adobe FrameMaker).

- **Conversion tools**

  Conversion tools convert documents from one format to another automatically. Typically, they convert printed documents to online formats (for example, Quadralay WebWorks Publisher). Many are designed to work with desktop publishing applications.

- **Content management systems**

  Content management systems (CMS) are used to manage large amounts of single-source content. Typically, they are database-driven tools that catalog information by element, not just document (for example, Arbortext Epic Editor and Documentum).

When evaluating development tools, look for authoring tools, conversion tools, and content management systems designed to work with each other. Often, tools that work particularly well together are produced by the same manufacturer. Sometimes, they are produced by different manufacturers. Some manufacturers allow you to test trial versions free of charge for a limited period of time.

Development tools designed to leverage the Standard Generalized Markup Language (SGML) or the Extensible Markup Language (XML) are particularly well suited to single sourcing because they separate content from format. These tools are normally very flexible and very powerful.

TIP For information about image processing tools and techniques, see "Optimizing images" on page 75.

About authoring tools

When evaluating authoring tools, look for industry-standard functions, such as the following:

- **Conditional text**

  Enables you to mark text or graphics for conditional display, then switch these sections on or off. For example, you can mark some sections of a document for online help only. You can then switch off these sections before printing the document.

  For details, see "Conditional text" on page 183.

- **Cross-referencing**

  Enables you to link elements electronically, so you can automatically generate dynamic cross-references, tables of contents, and indexes.

- **Large documents**

  Enables you to embed smaller documents in larger documents. For example, you can embed chapter documents in book documents. The best tools reference, rather than copy, small documents into large documents. They also enable you to reference charts and images into documents.

- **Page layout**

  Provides you with sets of templates for print or online documents. In more sophisticated programs, content elements are linked to format styles (for example, SGML elements are linked to paragraph styles, which are themselves linked to character styles).

- **Word processing**

  Enables you to enter document content elements (for example, text, figures, tables, and so on) from a graphical user interface. Normally, "what you see is what you get" (WYSIWYG). You can see your print or online output as you enter your input.

Single-source authoring tools are often desktop publishing applications.

TIP Adobe FrameMaker is a very popular desktop publishing application for large documents. The FrameMaker+SGML version of the product enables you to use Standard Generalized Markup Language (SGML) elements and DocBook templates when developing single-source content.

About conversion tools

When evaluating conversion tools, look for industry-standard functions, such as the following:

- **Automatic conversion**

  Converts source documents from one format to other formats automatically. Format elements or styles in the single-source document are mapped to format styles in target documents.

- **Conditional text**

  Interprets conditional text in source documents. For example, if you have marked some sections of your single-source document for print only, the sections do not appear in generated online help output. For details, see "Conditional text" on page 183.

- **Cross-referencing**

  Converts electronic cross-references, tables of contents, and indexes in source documents to hyperlinks. For example, cross-references with page numbers in a printed manual might be converted to clickable hyperlinks in an online help system.

- **Large documents**

  Converts large source documents that contain referenced documents. For example, you can convert a printed manual with chapters and sections to an online help system with sections and subsections. You can also convert sections of the printed manual into stand-alone files in the online help system.

- **Multiple document formats**

  Converts source documents to more than one format. For example, you can convert a printed manual to Microsoft HTML Help, Microsoft WinHelp, Oracle Help for Java, Sun JavaHelp, and so on.

- **Page layout**

  Provides sophisticated output templates you can customize. For example, you can change HyperText Markup Language (HTML) mappings and add Cascading Style Sheets (CSS).

Conversion tools often supplement desktop publishing applications.

TIP Adobe Acrobat and Quadralay WebWorks Publisher are two popular tools used to convert FrameMaker source documents into online formats.

About content management systems

Content management systems (CMS) are used to manage large amounts of single-source content. Because they catalog information by element rather than by document, these tools enable you to easily find and re-use the content modules you need. As a rule, these tools are database-driven.

When evaluating content management systems, look for industry-standard functions, such as the following:

- **Categorization**

  Categorizes document elements in a way that shows hierarchies and relationships. This function is analogous to a document index.

- **Searching**

  Integrates search engines (for example, the Verity Search Engine) that enable you to search documents by context as well as by keyword. For example, you can search a set of documents for a command name.

  For related guidelines, see "Search engines" on page 198.

- **Security**

  Tracks versions of the same document, so multiple writers can work on the same document without overwriting each other's work.

- **Version control**

  Tracks versions of the same document, so you can access and re-use earlier versions of the same document.

- **Workflow**

  Tracks the workflow of documents to make sure they follow your development process. This tracking includes checkpoints (for example, technical review) and approvals.

Content management systems are often designed to manage information developed with popular desktop publishing applications.

TIP Arbortext Epic Editor is a popular content management system used to develop SGML and XML content. This tool is designed to work with Documentum. It can also be adapted for the Oracle Internet File System.

High-end content management systems serve as complete authoring, conversion, and management systems. These systems are often used to generate tremendous amounts of dynamic content for online periodicals.

Filenames

The larger your single sourcing project, the more important intuitive filenames become. To re-use many different files for many different purposes, develop input filenames that are content-based rather than document-based. When generating output filenames, generate filenames that clearly indicate the content of the files *and* their "parent" documents.

Types of filenames

In most single sourcing environments, you develop two types of filenames:

- **Input filenames (manual)**

  You develop filenames for input files manually. For example, you could use an authoring tool to name the files that make up a book.

  For more about authoring tools, see "About authoring tools" on page 189.

- **Output filenames (automatic)**

  You generate filenames for output files automatically. For example, you could use a conversion tool to transform section titles of a book into filenames for an online help system.

  For more about conversion tools, see "About conversion tools" on page 190.

In both cases, you develop filenames that indicate what information the modules contain, *and* how that information is presented. The best way to indicate how information is presented is to follow a filename syntax that signals what type of modular content the file contains. To do so, you can establish a filename syntax that mirrors your heading syntax.

TIP For more about heading syntax, see "Headings" on page 85.

Developing input filenames

For input files, develop filenames that make sense even when the files are re-used in other documents. Develop filenames based on the content of the files, rather than on their role in the parent document. Whenever possible, use the document title (for example, the chapter title) in the filename.

Change	To
`Chapter1.fm`	`AcmeProAbout.fm`
`Chapter2.fm`	`AcmeProInstalling.fm`
`Chapter3.fm`	`AcmeProConfiguring.fm`
`Chapter4.fm`	`AcmeProUsing.fm`
`AppendixA.fm`	`AcmeProTroubleshooting.fm`
`AppendixB.fm`	`AcmeProCommands.fm`
`Glossary.fm`	`AcmeProGlossary.fm`
	`MasterGlossary.fm`

CAUTION When developing filenames, you must observe the filenaming constraints of your authoring tools, conversion tools, content management systems, and network systems. As an extreme example, in a DOS environment, you must follow the "8.3" filenaming convention. That is, you may use no more than eight characters for the filename itself, and three characters for the filename extension. Most development tools have less draconian filename constraints. Nevertheless, to avoid rework, check for any constraints on filename length *before* you begin naming files.

Generating output filenames

For output files, generate filenames that clearly indicate the content of the files *and* their parent documents. One easy way to indicate "parentage" in filenames is to generate prefixes based on module type.

Filename prefixes are especially helpful when you generate online help files based on sections and subsections in printed documents. In this scenario, you could convert hundreds of sections and subsections into files, all of which appear in your directory alphabetically. If you generate filename prefixes based on content type, you can easily locate the specific files you need. For example, filename prefixes might help you troubleshoot output problems in specific files, build a JavaScript table of contents for your online help, and so on.

Generating intuitive output filenames is easier than you might think. For example, if you are using Adobe FrameMaker as an authoring tool, and Quadralay WebWorks Publisher as a conversion tool, you can add filename markers to each heading of your input document. If you add filename prefixes to the filename makers in FrameMaker input files, WebWorks Publisher can automatically generate output files with the same filename prefixes.

Module type	**Filename prefix**
Definition list	`dl_`*`FileName`*`.html`
Glossary	`gl_`*`FileName`*`.html`
Procedure	`pd_`*`FileName`*`.html`
Process	`ps_`*`FileName`*`.html`
Topic	`to_`*`FileName`*`.html`
Troubleshooting scenario	`tr_`*`FileName`*`.html`

CAUTION When generating filenames, you must observe the filename constraints of your conversion and management tools, as well as your network systems. For example, WebWorks Publisher can generate filenames that are too long for some content management systems.

Localization

Localization involves document translation and customization for local markets. Although document localization is good for users, it also increases the complexity of your information development tasks. Fortunately, single sourcing can save you a tremendous amount of time and money when localizing documents. In fact, localization costs alone constitute an excellent argument for single sourcing.

Types of localization

Localization usually involves two separate but related activities:

- **Translation**

  Translating documents from one language to another. For example, you might author a document in English, then hire an outside agency to translate the document into Japanese.

  To find out how to leverage single sourcing when translating documents, see "Translating modular content" on page 197.

- **Customization**

  Customizing document content to match local versions of the product being documented. For example, the English and Japanese software bundles for your product might have different filenames.

  To find out how to assemble customized documents from a single source, see "Customizing modular content" on page 197.

If you address localization requirements up-front in your single sourcing process, you can save time and money.

TIP For more ways to use the single sourcing method to save time and money, see "Saving time and money" on page 8.

Types of translation

Essentially, there are three ways to translate documents:

- **Human translation**

  Translating texts manually. This method requires human translators who are fluent in the source language, and are native speakers of the target language. This method is used for documents that are not revised frequently.

- **Machine translation**

  Translating texts automatically. This method requires controlled language in original texts, and extensive post-translation editing by humans. This method is used by military and industrial organizations that are large and disciplined enough to leverage economies of scale.

  SYSTRAN is an example of machine translation software that is used by the U.S. Department of Defense, the Commission of European Communities, and large multinational corporations.

- **Translation memory**

  Translating texts semi-automatically. This method includes modifiable bilingual glossaries and "fuzzy memory." Fuzzy memory compares current texts with previous translations, allowing human translators to accept, reject, or edit those translations. This method requires almost no post-translation editing. It is used for publications that are revised frequently.

  TRADOS is an example of translation memory software that is used by many hardware and software manufacturers.

Of these three translation methods, machine translation (automatic) and translation memory (semi-automatic) are best suited to single sourcing. Like single sourcing, these translation methods eliminate duplicate work.

TIP SYSTRAN software is integrated into AltaVista Translation, or "Babel Fish." For details, see "Real-Time Machine Translation on the Internet" by Kurt Ament. This article appeared in the May 1998 issue of *Intercom*, the magazine of the Society for Technical Communication (STC).

Translating modular content

To maximize your translation dollars, you can translate modular content *after* it is edited but *before* it is assembled into documents. For example, if you have developed a printed manual in Adobe FrameMaker, you might translate this content into Japanese before you convert it into online help with Quadralay WebWorks Publisher. You would translate the same content once rather than twice. In other words, you would cut your translation bill in half, and reduce the chance of translation errors.

CAUTION When translating modular content, avoid translating filenames. The same filenames can be used by authoring tools and conversion tools to assemble documents in many different formats and languages. If you change these filenames, you cause yourself extra work for no reason.

For example, you might use filename markers in FrameMaker to generate HTML files in WebWorks Publisher. You might then reference the filenames in a JavaScript table of contents you build for the HTML files. If you were to translate the filenames, you would have to rebuild the JavaScript table of contents, filename by filename, for every language. In effect, you would have paid translators to waste your time.

For related guidelines, see "Filenames" on page 192.

Customizing modular content

To assemble customized documents from a single source, you can use conditional text to separate customized elements from generic elements. For example, if you know that the executable files for your product have different names and paths in the English- and Japanese-language versions of your product, you might set up conditional text styles named `EnglishOnly` and `JapaneseOnly`. To assemble the English-language version, you could switch on `EnglishOnly`, and switch off `JapaneseOnly`. To assemble the Japanese version, you could do the opposite.

TIP For more about conditional text, see "Conditional text" on page 183.

Search engines

Search engines enable users to search documents by title, header, text, and index. To maximize the effectiveness of your search engine, integrate search and index functions.

Types of search engines

There are two types of search engines:

- **Stand-alone**

  Stand-alone search engines (for example, the Verity Search Engine) can be used by themselves. Typically, these search engines are used to search many different documents simultaneously.

- **Integrated**

  Integrated search engines (for example, the search function in Microsoft HTML Help) are built into other software applications. Typically, these search engines are associated with one document only.

Sometimes, the distinction between stand-alone search engines and integrated search engines becomes blurred. For example, some content management systems integrate stand-alone search engines.

TIP For a list of standard functions found in content management systems, see "About content management systems" on page 191.

Types of searches

You can use search engines to conduct many different types of searches:

- Title
- Header
- Full text
- Index

To make your documents as accessible as possible, develop an integrated indexing and searching strategy. That is, index your searches, and search your indexes.

Refining searches

To control search results, refine your documents as follows:

- **Refining titles**

  Title searches enable users to search documents by title. For example, a search for the word "printing" would return a list of all documents with "printing" in their titles. To improve title search results, use very descriptive headings in your documents.

  For related guidelines, see "Headings" on page 85.

- **Refining headers**

  Header searches enable users to search documents by keywords. Keywords are located in document headers, which are visible to search engines only. For example, if you add the keywords "printing document" to a document header, a search that uses this exact keyword combination would find the document. However, a search that used "documents, printing" (plural) might not. To improve header search results, add very descriptive keywords to your document headers.

- **Refining text**

  Full-text searches enable users to search the text of documents. For example, a search for "printing" would return a list of all documents that contain the word "printing." Normally, full-text searches rank results by "hits." All too often, full-text searches provide false positives. Internet search engines showcase this problem. To improve full-text search results, use very precise language in your modules, and organize documents by module type.

  For related guidelines, see Chapter 4, "Configuring language."

- **Refining indexes**

  Indexes enable users to see hierarchies and relationships between topics, procedures, and so on. Because indexes are much more sophisticated than title, header, or full-text searches, you should look for ways to re-use indexes in your searches. If possible, enable your search engine to search indexes, not just documents. To improve index search results, follow a strict and sensible indexing strategy.

  For related guidelines, see "Indexes" on page 92.

Variables

Variables are definitions for dynamic text. Most authoring and conversion tools used to build single-source documents enable you to create variables when you expect text to change. For each variable, you create a name and a definition. You then insert the variable instead of text throughout your documents. When you change the variable definition, all occurrences of the variable in your documents are updated with the new definition.

Types of variables

Most authoring tools have two types of variables:

- **System variables**

  Authoring tools define system variables (for example, the current date, chapter titles, page numbers, and so on).

- **User variables**

  You define user variables (for example, product names, product versions, and document titles).

You can combine these two kinds of variables to build very complex variables. For example, you could set up a variable for running headers that automatically repeat chapter titles, which in turn contain product names defined in variables.

Where to use variables

You can define variables in your authoring tool or your conversion tool:

- **Authoring tool (content)**

  Define content-related variables (for example, product names) in your authoring tool. These variables do *not* change from format to format.

  For related guidelines, see "About authoring tools" on page 189.

- **Conversion tool (format)**

  Define format-related variables (for example, titles for converted online help documents) in your conversion tool. These variables do change from format to format.

  For related guidelines, see "About conversion tools" on page 190.

Building variables

For variable names, use a consistent syntax that enables you to create a large library of variables. Make sure the variable names are easy for any writer to find and understand. Use system variables and user variables as "building blocks" when creating other variables. For example, when you create variables for document titles, you could re-use system variables for typographical styles. And you could re-use user variables for product names and versions.

Name	**Definition**	**Output**
`Prod_Name`	`AcmePro`	AcmePro
`Prod_Version`	`7.0`	7.0
`Title_InstallGuide`	<italic> Installing <$Prod_Name> <$Prod_Version> <DefaultParaFont>	*Installing AcmePro 7.0*
`Title_UserGuide`	<italic> Using <$Prod_Name> <$Prod_Version> <DefaultParaFont>	*Using AcmePro 7.0*

Maintaining variables

To keep variables consistent and current, maintain them in a single-source document shared by all writers, but coordinated by one writer or editor:

1. **Compile**
   Coordinator builds one document that contains the latest variables.
2. **Update**
   Writers submit new variable names and definitions, as needed.
3. **Distribute**
   Coordinator distributes the variables document, as it is updated.

TIP For more about teamwork, see "Organizing smart teams" on page 21.

GLOSSARY

assembly

Process of transforming modular content into a specific document format for a specific audience and purpose. This process involves document organization, linking, and conversion.

See also conversion; linking; module; primary module; re-assembly; secondary module; single sourcing.

authoring tool

Software application used to author complex documents (for example, Adobe FrameMaker). Typically, the application is a desktop publishing application that enables you to develop print and online documents from a single source. Many such applications are designed to work with popular document conversion tools.

See also CMS; conditional text; conversion tool; variable.

Cascading Style Sheet

See CSS.

chunking

Process of organizing content into stand-alone modules, based on the type of information being presented. For example, you might use descriptive text to explain what a product does. And you might use procedures to explain how to operate the product.

See also labeling; linking; module.

CMS

content management system. Software application used to manage large amounts of single-source content (for example, Arbortext Epic Editor). Typically, the application is a database-driven tool that catalogs information by element, not just document. Many such applications are designed to work with popular document authoring and conversion tools.

See also authoring tool; conversion tool.

conditional text

Function in many authoring and conversion tools that enables you to mark textual or graphical elements for conditional display, then switch the elements on or off, as needed. For example, you could mark some sections of a document for online help only, then switch them off before printing the document.

See also authoring tool; conversion tool.

content management system

See CMS.

conversion

Process of mechanically transforming documents from one format to another format (for example, from SGML to HTML). The goal of this process is to ensure consistency of identical content in different document formats. Unlike repurposing, this process is mechanical. The process is best performed by machines.

See also assembly; conversion tool; HTML; re-assembly; repurposing; SGML.

conversion tool

Software applications used to automatically convert documents from one format to another format (for example, Quadralay WebWorks Publisher). Typically, the application converts print documents to online formats. Many such applications are designed to work with popular authoring tools.

See also authoring tool; CMS; conditional text; conversion; variable.

CSS

Cascading Style Sheet. Extension to HTML that allows you to define styles (for example, font family) for certain elements in HTML documents. You can include style information in the header of an HTML file, or in a separate file that can then be shared by multiple HTML files. You can also combine style sheets with JavaScript to produce DHTML. The style sheets are to HTML what XSL is to XML.

See also DHTML; HTML; JavaScript; XML; XSL.

development tool

See authoring tool; CMS; conversion tool.

DHTML

Dynamic HTML. Extension to HTML developed by Microsoft Corporation to provide developers with greater control over page layout. This extension enables you to produce web pages that change and interact with users, without communicating with a server. To produce these effects, you combine CSS and JavaScript.

See also CSS; HTML; JavaScript.

DocBook

SGML-based DTD used to write structured documents with SGML or XML (for example, with Adobe FrameMaker+SGML). This DTD is particularly well-suited to hardware and software documentation.

See also DTD; element; SGML; tag; XML.

document type definition

See DTD.

DTD

document type definition. Definition of a document type in SGML or XML. This definition consists of a set of markup tags and their interpretation (for example, DocBook).

See also DocBook; SGML; XML.

Dynamic HTML

See DHTML.

element

Node in an SGML, XML, or HTML document tree that defines the hierarchical structure of a document. Most of these nodes have start and end tags, which contain document content (for example, paragraphs or lists). Ideally, single-source documents are written at this content level, rather than the format level.

See also DocBook; HTML; modular writing; SGML; tag; XML.

Extensible Markup Language

See XML.

Extensible Stylesheet Language

See XSL.

GIF

Graphics Interchange Format. Standard defined by CompuServe Incorporated for digitized images compressed with the LZW algorithm. This standard is used in HTML pages for images that contain line drawings or text.

See also HTML; JPEG; LZW compression; TIFF.

Graphics Interchange Format

See GIF.

HTML

HyperText Markup Language. Hypertext document format based on SGML. This format is used to create pages you can view with a web browser or in online help systems. You can use conversion tools to generate pages in this format from SGML or XML documents. Think of this format as the grandchild of SGML, and the child of XML.

See also conversion; CSS; DHTML; element; GIF; JPEG; SGML; tag; XML; XSL.

HyperText Markup Language

See HTML.

information mapping

Method for analyzing, organizing, and presenting information developed by Robert E. Horn, founder of Information Mapping, Inc. "Based on research into how the human mind actually reads, processes, remembers, and retrieves information," this method "enables authors to break complex information into its most basic elements and then present those elements optimally for readers." As such, the method is ideal for single sourcing.

See also modular writing.

JavaScript

Simple, cross-platform scripting language developed by Netscape Communications Corporation for use on the Internet. This language runs as a server-side scripting language, as an embedded language in server-parsed HTML, and as an embedded language run in web browsers. You can combine this language with CSS to create DHTML effects.

See also CSS; DHTML; HTML.

Joint Photographic Experts Group

See JPEG.

JPEG

Joint Photographic Experts Group. 1. Standard image format used to compress full-color or grayscale images of realistic objects in HTML pages (for example, digital photographs). This format does not work well with non-realistic images (for example, screenshots or line art). 2. Committee that designed the standard image compression algorithm that bears its name.

See also GIF; HTML; TIFF.

labeling

Process of adding headings to textual modules (for example, procedures and topics), and adding captions to visual modules (for example, figures and tables). Headings and captions should follow a standard, context-independent syntax to identify modules so they make sense in any document or format. For example, you might label superprocedures with gerunds, and you might label subprocedures with infinitives.

See also chunking; linking.

Lempel-Ziv Welch compression

See LZW compression.

linear writing

Document-based writing method. Using this method, you organize information into sequential hierarchies, based on the structure of documents (for example, chapters, sections, and subsections). You design documents to be read sequentially, from beginning to end. Because content is tied to one specific document format, information is almost impossible to re-use in other formats.

See also modular writing.

linking

Process of cross-referencing labeled modules. For example, you might cross-reference each step in a superprocedure to its related subprocedure, and each subprocedure to related procedures, topics, and references. Likewise, you might electronically mark all topics, procedures, and definition lists, then generate a unique table of contents and index for each document you assemble.

See also assembly; chunking; labeling; re-assembly.

LZW compression

Lempel-Ziv Welch compression. Algorithm used to reduce the size of files, especially for archival or transmission. This algorithm was designed by Terry Welch for high-performance disk controllers. It is now used in GIF images.

See also GIF.

management tool

See CMS.

mapping

See information mapping; linking.

master glossary

Glossary shared by related documents. This shared glossary enables users to find definitions to technical terms in document sets. It also enables writers to reduce redundancy and inconsistency in terms and definitions.

See also master index.

master index

Index shared by related documents. This shared index enables users to quickly find information in document sets. It also helps writers to reduce redundancy and inconsistency in document sets.

See also master glossary.

modular writing

Element-based writing method. Using this method, you organize information into stand-alone content modules, based on document elements (for example, paragraphs and lists). You design content modules to be read in any context. Because it separates content from format, this writing method results in content that is easy to re-use. This writing method drives the single sourcing process.

See also element; information mapping; linear writing; module; single sourcing.

module

Stand-alone chunk of information (for example, a procedure or a topic) that answers one basic question: who, what, when, where, why, or how. You assemble these chunks into documents.

See also assembly; chunking; modular writing; primary module; secondary module.

PDF

Portable Document Format. Native file format for Adobe Systems Acrobat. This file format represents documents in a manner that is independent of the original application software, hardware, and operating system used to create those documents. Technical writers often use this format for print-ready electronic books that can be viewed in web browsers.

Portable Document Format

See PDF.

primary module

Stand-alone chunk of information you assemble as a document section. Each document section is comprised of one stand-alone chunk (for example, a definition list, glossary, procedure, process, topic, or troubleshooting scenario). Sections can contain secondary modules as well.

See also assembly; module; secondary module.

re-assembly

Process of transforming previously assembled modular content into a new document format for a new audience and purpose. Like assembly, this process involves document organization, linking, and conversion.

See also assembly; conversion; linking; repurposing; single sourcing.

repurposing

Process of transforming modular content developed in one document format (for example, a printed manual) so that it makes sense in another document format (for example, an online help system). The main goal of this process is to ensure the usability of identical content in different document formats. Unlike conversion, this process is a cognitive process best performed by humans.

See also conversion; re-assembly; single sourcing.

secondary module

"Helper" module (for example, an example, a figure, an itemized list, a note, or a table) you integrate into a primary module.

See also assembly; module; secondary module.

SGML

Standard Generalized Markup Language. Generic markup language used to represent documents. This language is an international standard that describes the relationship between document content and its structure. The language allows you to share and re-use documents across applications and operating systems in an open, vendor-neutral format. You can convert this information into other formats (for example, HTML) with document conversion tools. Think of this language as the parent of XML, and the grandparent of HTML.

See also conversion; conversion tool; DocBook; DTD; element; HTML; tag; XML; XSL.

single sourcing

Method for systematically re-using information. With this method, you develop modular content in one source document or database. You then assemble that content into different document formats for different audiences and purposes.

See also assembly; modular writing; re-assembly; repurposing.

Standard Generalized Markup Language

See SGML.

tag

Token representing the beginning or end of an element in an SGML, XML, or HTML document. The token is normally enclosed in angle brackets. For example, `<Para>` is used to mark the beginning of paragraphs in SGML DocBook documents.

See also DocBook; element; HTML; SGML; XML.

Tagged Image File Format

See TIFF.

TIFF

Tagged Image File Format. File format used for still-image bitmaps, stored in tagged fields. Application programs can use the tags to accept or ignore fields, depending on their capabilities. This format is normally used for printed documents.

See also GIF; JPEG.

variable

Dynamic text element. Most authoring and conversion tools enable you to define dynamic text when you expect text to change. For each dynamic text element, you create a name and a definition. You then insert the element instead of text throughout your documents. When you change the element definition, all occurrences of the element in your documents are updated with the new definition.

See also authoring tool; conversion tool.

XML

Extensible Markup Language. Dialect of SGML suitable for use on the Internet. This dialect is the basis of many database-driven single sourcing applications. You can use XSL to convert documents in this dialect to HTML. Think of this dialect as the child of SGML, and the parent of HTML.

See also CSS; DocBook; DTD; element; HTML; SGML; tag; XSL.

XSL

Extensible Stylesheet Language. Standard for defining XML style sheets. You can use these style sheets to transform XML content into HTML format. These style sheets are to XML what CSS style sheets are to HTML.

See also CSS; HTML; SGML; XML.

INDEX

B

C

D

H

I

J

K

L

M

N

O

P

U

V